# Six-Minute Solutions
## for Mechanical PE Exam
## Thermal and Fluids Systems Problems

### Second Edition

Daniel C. Deckler, PhD, PE

Professional Publications, Inc. • Belmont, California

## Benefit by Registering This Book with PPI

- Get book updates and corrections.
- Hear the latest exam news.
- Obtain exclusive exam tips and strategies.
- Receive special discounts.

Register your book at **ppi2pass.com/register**.

## Report Errors and View Corrections for This Book

PPI is grateful to every reader who notifies us of a possible error. Your feedback allows us to improve the quality and accuracy of our products. You can report errata and view corrections at **ppi2pass.com/errata**.

**SIX-MINUTE SOLUTIONS FOR MECHANICAL PE EXAM THERMAL AND FLUIDS SYSTEMS PROBLEMS**
Second Edition

Current printing of this edition: 8

### Printing History

| date | edition number | printing number | update |
|---|---|---|---|
| Dec 2012 | 2 | 6 | Minor corrections. |
| Jan 2015 | 2 | 7 | Minor corrections. Copyright update. |
| Nov 2015 | 2 | 8 | Minor cover updates. |

© 2015 Professional Publications, Inc. All rights reserved.

All content is copyrighted by Professional Publications, Inc. (PPI). No part, either text or image, may be used for any purpose other than personal use. Reproduction, modification, storage in a retrieval system or retransmission, in any form or by any means, electronic, mechanical, or otherwise, for reasons other than personal use, without prior written permission from the publisher is strictly prohibited. For written permission, contact PPI at permissions@ppi2pass.com.

Printed in the United States of America.

PPI
1250 Fifth Avenue
Belmont, CA 94002
(650) 593-9119
ppi2pass.com

ISBN: 978-1-59126-147-6

Library of Congress Control Number: 2008938164

F E D C B A

# Table of Contents

**ABOUT THE AUTHOR** . . . . . . . . . . . . . . . . . . . . iv

**PREFACE AND ACKNOWLEDGMENTS** . . . . . . . . . . . . v

**INTRODUCTION**
    Exam Format . . . . . . . . . . . . . . . . . . . . . . . vii
    This Book's Organization . . . . . . . . . . . . . . . . . vii
    How to Use This Book . . . . . . . . . . . . . . . . . . vii

**REFERENCES** . . . . . . . . . . . . . . . . . . . . . . . . ix

**NOMENCLATURE** . . . . . . . . . . . . . . . . . . . . . xi

**BREADTH PROBLEMS**
    Hydraulics and Fluids . . . . . . . . . . . . . . . . . . . 1
    Energy/Power Systems . . . . . . . . . . . . . . . . . . 3

**DEPTH PROBLEMS**
    Principles . . . . . . . . . . . . . . . . . . . . . . . . . 5
    Equipment . . . . . . . . . . . . . . . . . . . . . . . . 11
    Systems . . . . . . . . . . . . . . . . . . . . . . . . . . 16
    Codes and Standards . . . . . . . . . . . . . . . . . . . 17

**BREADTH SOLUTIONS**
    Hydraulics and Fluids . . . . . . . . . . . . . . . . . . . 19
    Energy/Power Systems . . . . . . . . . . . . . . . . . . 25

**DEPTH SOLUTIONS**
    Principles . . . . . . . . . . . . . . . . . . . . . . . . . 29
    Equipment . . . . . . . . . . . . . . . . . . . . . . . . 45
    Systems . . . . . . . . . . . . . . . . . . . . . . . . . . 55
    Codes and Standards . . . . . . . . . . . . . . . . . . . 61

# About the Author

Daniel C. Deckler was born in Tiffin, Ohio, and raised in various parts of Ohio. He received his bachelor's degree in mechanical engineering from Ohio Northern University, and received both his master's degree in mechanical engineering and his PhD in engineering from the University of Akron. He received his professional engineer's license in 1999.

Dr. Deckler is currently a professor of engineering at the University of Akron Wayne College, where he teaches courses in mechanical, electrical, and civil engineering. Before coming to the University of Akron, he was an engineer for Rockwell International in Downey, California, the Timken Company in Canton, Ohio, and the Loral Corporation (formerly Goodyear Aerospace) in Akron, Ohio. With these companies he worked on interesting projects such as the space shuttle and space station, bearing applications, and anti-submarine warfare systems, respectively.

Dr. Deckler lives in Canton, Ohio. When not working, he can be found enjoying the outdoors, be it golfing, cycling, running or training for triathlons, or hunting.

# Preface and Acknowledgments

The Principles and Practice of Engineering examination (or PE exam for short) for mechanical engineers, prepared by the National Council of Examiners for Engineering (NCEES), is developed from sample problems submitted by educators and professional engineers representing consulting, government, and industry. PE exams are designed to test examinees' understanding of both theoretical and practical engineering concepts. Problems from past exams are not available from NCEES or any other source. However, NCEES does give the general subject areas covered on the exam.

The topics covered in *Six-Minute Solutions for Mechanical PE Exam Thermal and Fluids Systems Problems* are the same subject areas identified by NCEES for the thermal and fluids systems depth module of the mechanical PE exam. These problems cover fluid mechanics, heat transfer, compressible and incompressible flow, thermodynamic cycles, pumps, compressors, turbines, and more.

Many successful examinees agree that, along with learning to retrieve information rapidly from reference materials, working practice problems instills a valuable level of confidence. The problems presented here originate from both professional practice and instructional experience. This book serves as a companion to the *Mechanical Engineering Reference Manual* and is designed primarily to help you remember what you have already learned. I have purposefully left "shortcuts" out of the problem solutions. Instead, I have focused on solving the problems from fundamental concepts. Solving in this manner might take a few more minutes, but it eliminates the need to remember special cases. One only has to recall a few basic ideas that are easily remembered or located. In addition, for several of the problems I have shown multiple methods of solution, to further strengthen your review.

The problems presented in this book are representative of the type and difficulty of those you will encounter on the PE exam. The book's problems are both conceptual and practical, and they are written to provide varying levels of difficulty. Though you probably won't encounter exam problems on exactly the same situations, reviewing these problems and solutions will increase your familiarity with the exam problems' form, content, and solution methods. This preparation will help you considerably during the exam.

Problems and solutions have been carefully prepared and reviewed to ensure that they are appropriate, understandable, and correctly solved. If you find errors or discover alternative, more efficient methods of solving a problem, please bring it to PPI's attention so that your suggestions can be incorporated into future editions. You can report errors and keep up with the changes made to this book, as well as changes to the exam, by going to PPI's website at **ppi2pass.com/errata**.

For the second edition, the problems have been reorganized. The order of topics in this book now matches the organization of subject areas as given in the current NCEES specifications for the mechnical PE exam, which became effective in October 2008. We hope this will help you focus your study to the needs of the current exam.

This book would not have been possible without the contributions of the many people who supported me through several years of writing and review.

A huge thank you goes to Jan Deckler. She is an engineer at a refinery, and many of the problems you see in this book are the direct result of her experience. So, whenever you see a problem about a refinery, you will now know where it came from. Another thank you goes to all my former engineering coworkers who were kind enough to take me "under their wings" and teach me what they knew when I was a beginning engineer. Believe it or not, they were an inspiration for several problems so many years later. Finally, I would like to thank my former and current students who are interested in real-life applications of the subjects I teach. They are always making me reach into my memory and pull out useful problems that I had to solve during my days in industry and as a consultant. Several of these problems have found their way into this book.

From PPI, I recognize Aline Magee, who started me along this journey; Sarah Hubbard, who helped me along the way; and Scott Marley, who persisted with me to the completion of this book. Scott is a fiend about deadlines, and there were times when I found him infuriating. Scott, however, has the patience of a saint and stuck with me despite my failings, and for that I owe him a great deal of thanks. Thank you, Scott, this book would never have been completed without your dedication. To David Bostain, Matt Gordon, and Muamba Wanzala, who reviewed and helped polish the manuscript, and to the rest of the editorial and production staff at PPI, especially Miriam Hanes, Kate Hayes, and Amy Schwertman, thank you all. I very much appreciate your hard work.

To all of you who read this book, it is my greatest hope that you find it useful. Good luck.

Daniel C. Deckler, PhD, PE

# Introduction

## EXAM FORMAT

The Principles and Practice of Engineering examination (PE exam) in mechanical engineering is an 8-hour exam divided into a morning and an afternoon session. The morning session is known as the "breadth" exam and the afternoon is known as the "depth" exam.

The morning session includes 40 problems from all three areas of mechanical engineering: HVAC and refrigeration, mechanical systems and materials, and thermal and fluids systems. The three areas are roughly equally represented. As the "breadth" designation implies, morning session problems are general in nature and wide-ranging in scope.

The afternoon session allows the examinee to select a "depth" exam module from one of the three subdisciplines. The 40 problems included in the afternoon session require more specialized knowledge than those in the morning session.

All problems from both the morning and afternoon sessions are multiple choice. They include a problem statement with all required defining information, followed by four logical choices. Only one of the four options is correct. Problems are generally self-contained and independent, so an incorrect choice on one problem typically will not carry over to subsequent problems.

Topics and the approximate distribution of problems on the afternoon session of the mechanical thermal and fluids systems exam are as follows.

### Principles: about 45% of exam problems

- Properties of materials
- Fluid mechanics: compressible fluids, incompressible fluids
- Principles of heat transfer
- Principles of mass balance
- Thermodynamics: cycles, properties, energy balances, combustion
- Related principles: strength of materials, fatigue theory, statics and dynamics, stress analysis, psychrometrics, welding, safety, quality control and quality assurance

### Applications: about 55% of exam problems

- Equipment: pumps, turbines, compressors, fans, blowers, boilers, steam generators, engines, drive trains, pressure vessels, heat exchangers, condensers, feedwater heaters, cooling towers, valves and other control devices
- Systems: power hydraulics, pneumatic systems, fluid distribution, power conversion, energy recovery, cooling and heating, power cycles
- Codes and standards

For further information and tips on how to prepare for the mechanical engineering PE exam, consult the *Mechanical Engineering Reference Manual* or the PPI website, **ppi2pass.com/mefaq**.

## THIS BOOK'S ORGANIZATION

*Six-Minute Solutions for Mechanical PE Exam Thermal and Fluids Systems Problems* is organized into two sections. The first section, Breadth Problems, presents 20 problems in thermal and fluids systems of the type that would be expected in the morning part of the mechanical engineering PE exam. The second section, Depth Problems, presents 65 problems typical of the afternoon part of this exam. Each of these two sections of the book is further subdivided into more specific topic areas covered by the thermal and fluids systems exam.

Most of the problems are quantitative, requiring calculations to arrive at a correct solution. A few are non-quantitative. Some problems will require a little more than 6 minutes to answer and others a little less. On average, you should expect to complete 80 problems in 480 minutes (8 hours), or spend 6 minutes per problem.

*Six-Minute Solutions for Mechanical PE Exam Thermal and Fluids Systems Problems* does not include problems related directly to HVAC and refrigeration, nor to mechanical systems and materials, although problems from these subdisciplines will be included in the thermal and fluids systems exam, particularly in the morning session. Other books in the *Six-Minute Solutions* series provide problems for review in these areas.

## HOW TO USE THIS BOOK

Each problem statement in this book, with its supporting information and answer choices, is presented in the same format as the problems encountered on the PE exam. The solutions explain step by step how the answer is logically derived, to help you follow the reasoning and to provide examples of how you may want to approach your solutions as you take the PE exam.

Each problem includes a hint to provide direction in solving the problem. In addition to the correct solution, you will find an explanation of the faulty reasoning leading to the three incorrect answer choices. The incorrect answers are chosen to show some common mistakes made when solving each type of problem. These may be simple mathematical errors, such as failing to square a term in an equation, or more serious errors, such as using the wrong equation.

To optimize your study time and obtain the maximum benefit from the practice problems, consider the following suggestions.

1. Complete an overall review of the problems and identify the subjects that you are least familiar with. Work a few of these problems to assess your general understanding of the subjects and to identify your strengths and weaknesses.

2. Locate and organize relevant resource materials. (See the references section of this book as a starting point.) As you work problems, some of these resources will emerge as more useful to you than others. These are what you will want to have on hand when taking the PE exam.

3. Work the problems in one subject area at a time, starting with the subject areas that you have the most difficulty with.

4. When possible, work problems without using the hint. Always attempt your own solution before looking at the solutions provided in the book. Use the solutions to check your work or to provide guidance in finding solutions to the more difficult problems. Use the incorrect solutions to help identify pitfalls and to develop strategies to avoid them.

5. Use each subject area's solutions as a guide to understanding general problem-solving approaches. Although problems identical to those presented in this book will not be encountered on the PE exam, the approach to solving problems will be the same.

Solutions presented for each example problem may represent only one of several methods for obtaining a correct answer. Although we have tried to prepare problems with unique solutions, alternative problem-solving methods may occasionally produce a different, but nonetheless appropriate, answer.

# References

The minimum recommended library for the mechanical PE exam is PPI's *Mechanical Engineering Reference Manual*. You may also find the following references helpful in completing some of the problems in *Six-Minute Solutions for Mechanical PE Exam Thermal and Fluids Systems Problems*.

American Petroleum Institute. API 570-1998: *Piping Inspection Code: Inspection, Repair, Alteration, and Rerating of In-service Piping Systems*.

_____. API 682-2004/ISO 21049: *Pumps—Shaft Sealing Systems for Centrifugal and Rotary Pumps*.

American Society of Mechanical Engineers (ASME). B31.3-2006: *Process Piping*.

_____. *Boiler and Pressure Vessel Code, 2007 Edition*.

Keenan, Joseph H., et al., *Steam Tables: Thermodynamic Properties of Water Including Vapor, Liquid, and Solid Phases*. Krieger Publishing Company.

Shigley, Joseph E., et al., *Mechanical Engineering Design*. McGraw-Hill.

Volk, Michael W., *Pump Characteristics and Applications*. Marcel Dekker.

# Nomenclature

| | | | | | | | |
|---|---|---|---|---|---|---|---|
| $a$ | acceleration | ft/sec$^2$ | m/s$^2$ | $h_m$ | mass transfer coefficient | ft/sec | m/s |
| $a$ | speed of sound | ft/sec | m/s | $H$ | differential pressure | lbf/ft$^2$ | Pa |
| $A$ | area | ft$^2$ | m$^2$ | $H$ | heat dissipation | Btu-hr/bhp | n.a. |
| ACFM | actual cubic feet per minute | ft$^3$/min | n.a. | $i$ | effective rate of return | % | % |
| BHP | boiler or brake horsepower | hp | hp | $I$ | current | A | A |
| $c$ | clearance | ft | m | $I$ | mass moment of inertia | ft-lbf-sec$^2$ | m·N·s$^2$ |
| $c$ | distance from neutral axis to extreme fiber | ft | m | $J$ | polar area moment of inertia | ft$^4$ | m$^4$ |
| $c_p$ | specific heat | Btu/lbm-°F | J/kg·°C | $k$ | ratio of specific heats | – | – |
| $C$ | attachment factor | – | – | $k$ | thermal conductivity | Btu/hr-ft-°F | W/m·°C |
| $C$ | cost | $ | $ | $K$ | constant | – | – |
| $C$ | thermal capacitance | Btu/°F | J/°C | $K$ | minor loss coefficient | – | – |
| CR | circulation ratio | – | – | $K_d$ | density factor | – | – |
| CR | corrosion rate | mil/yr | n.a. | $L$ | flow resistance | liquid ohm | liquid ohm |
| $d$ | diameter | ft | m | $L$ | length | ft | m |
| $d$ | moment arm | ft | m | $L$ | life | yr | yr |
| $D$ | depreciation | $ | $ | $m$ | mass | lbm | kg |
| $E$ | efficiency | – | – | $\dot{m}$ | mass flow rate | lbm/hr | kg/h |
| $E$ | quality factor | – | – | $M$ | Mach number | – | – |
| $E$ | specific energy | ft-lbf/lbm | J/kg | $n$ | number of periods | – | – |
| $f$ | frequency | Hz | Hz | $n$ | rotational speed | rpm | rpm |
| $f$ | friction factor | – | – | $n_s$ | specific speed | – | – |
| $F$ | force, load | lbf | N | $N$ | number of poles | – | – |
| $g$ | gravitational acceleration | ft/sec$^2$ | m/s$^2$ | NI | net income | $ | $ |
| $g_c$ | gravitational constant | ft-lbm/lbf-sec$^2$ | n.a. | NPSHA | net positive suction head available | ft | m |
| $G$ | mass flow rate per unit area | lbm/ft$^2$-sec | kg/m$^2$·s | $p$ | pressure | lbf/ft$^2$ | Pa |
| $G(s)$ | transfer function | – | – | $P$ | profit | $ | $ |
| $h$ | convection or film coefficient | Btu/hr-ft$^2$-°F | W/m$^2$·°C | $P$ | present worth | $ | $ |
| $h$ | enthalpy | Btu/lbm | J/kg | $P$ | power | hp | W |
| $h$ | head, head loss, or height | ft | m | $P$ | wetted perimeter | ft | m |
| $h$ | weld depth | ft | m | $q$ | heat | Btu | J |

| | | | |
|---|---|---|---|
| $Q$ | heat flow rate, heat transfer rate | Btu/hr | W |
| $Q$ | flow rate | gal/sec | L/s |
| $r$ | radius | ft | m |
| $r$ | rate | % | % |
| $R$ | reaction | lbf | kg |
| $R$ | specific gas constant | ft-lbf/lbm-°R | J/kg·K |
| RCA | remaining corrosion allowance | ft | m |
| Re | Reynolds number | – | – |
| RF | rating factor | – | – |
| $R(s)$ | response function | – | – |
| $s$ | entropy | Btu/lbm-°F | J/kg·°C |
| $S$ | Sommerfeld number | – | – |
| $S$ | stress value | lbf/ft$^2$ | Pa |
| $S_n$ | expected salvage value in year $n$ | $ | $ |
| SCFM | standard cubic feet per minute | ft$^3$/min | n.a. |
| SG | specific gravity | – | – |
| Sl | Strouhal number | – | – |
| $t$ | thickness | ft | m |
| $t$ | time | sec | s |
| $T$ | taxes | $ | $ |
| $T$ | temperature | °F | °C |
| $T$ | tension | lbf | N |
| $T$ | torque | ft-lbf | N·m |
| TU | tower units | – | – |
| $U$ | overall heat transfer coefficient | Btu/hr-ft-°F | W/m$^2$·°C |
| v | velocity | ft/sec | m/s |
| $V$ | shear force | lbf | N |
| $V$ | voltage | V | V |
| $V$ | volume | ft$^3$ | m$^3$ |
| $w$ | weight per unit length | lbf/ft | N/m |
| $w$ | width | ft | m |
| $W$ | mass flow rate | lbm/sec | kg/s |
| $W$ | weight | lbf | N |
| $W$ | work | Btu/lbm | J/kg |
| WHP | water horsepower | hp | hp |
| $x$ | quality | – | – |
| $z$ | elevation | ft | m |

**Symbols**

| | | | |
|---|---|---|---|
| $\alpha$ | diffusivity | ft$^2$/sec | m$^2$/s |
| $\beta$ | angle of wrap | deg | deg |
| $\gamma$ | specific weight | lbf/ft$^3$ | n.a. |
| $\epsilon$ | fractional error | – | – |
| $\epsilon$ | specific roughness | ft | m |
| $\eta$ | efficiency | – | – |
| $\mu$ | absolute viscosity | lbf-sec/ft$^2$ | Pa·s |
| $\mu$ | coefficient of friction | – | – |
| $\nu$ | kinematic viscosity | ft$^2$/sec | m$^2$/s |
| $\rho$ | density | lbm/ft$^3$ | kg/m$^3$ |
| $\sigma$ | cavitation number | – | – |
| $\sigma$ | stress | lbf/ft$^2$ | Pa |
| $\tau$ | shear stress | lbf/ft$^2$ | Pa |
| $v$ | specific volume | ft$^3$/lbm | m$^3$/kg |
| $\phi$ | relative humidity | – | – |
| $\omega$ | angular velocity | rpm | rpm |
| $\omega$ | humidity ratio | – | – |

**Subscripts**

| | |
|---|---|
| $\phi$ | original, stagnation |
| $A$ | added (by pump) |
| add | additional |
| atm | atmospheric |
| ave | average |
| $b$ | base |
| $B$ | bottom |
| bend | bending |
| $c$ | compressible |
| comb | combustion |
| corr | corrected |
| cr | critical |
| $d$ | design, density |
| $D$ | discharge |
| $E$ | extracted (by turbine) |
| eq | equivalent |
| $f$ | fin, final, fluid, friction |
| $fg$ | fluid to gas (vaporization) |
| $g$ | gas |
| $h$ | hydraulic |
| $H$ | high |
| HHV | higher heating value |
| $i$ | incompressible, inflow, initial, inner |
| in | in, input |
| inj | injection |
| $L$ | left, low |
| $m$ | minor, motor |

| | |
|---|---|
| max | maximum |
| $n$ | nozzle |
| $o$ | outer |
| off | off-design |
| os | oversized |
| out | out, output |
| $p$ | pipe, pressure |
| $P$ | polar |
| $r$ | radial, reinforcement |
| $R$ | rated, right |
| req | required |
| $s$ | shell, specific, surface |
| $S$ | suction |
| sat | saturation |
| std | standard |
| $t$ | tangential, turbine |
| $T$ | tank |
| th | theoretical, thermal |
| $v$ | valve |
| vp | vapor pressure |
| wv | water vapor |
| $x$ | horizontal position |
| $y$ | yield, vertical position |
| $z$ | potential, position along $z$-axis |
| $\infty$ | at infinity |

# Breadth Problems

## HYDRAULICS AND FLUIDS

### PROBLEM 1

An F-14 Tomcat is flying at an altitude of 50,000 ft at 1200 mph. The temperature at this altitude is 393°R. If air is treated as an ideal gas, the Tomcat's Mach number is most nearly

    (A) 0.55
    (B) 1.6
    (C) 1.8
    (D) 2.1

Hint: Understand the definition of Mach number.

### PROBLEM 2

Using international standard atmosphere conditions, the maximum pressure on the nose of a plane flying at Mach 1 at sea level can be calculated using the compressible flow theory. Using incompressible flow theory will introduce an error that is most nearly

    (A) 10%
    (B) 21%
    (C) 27%
    (D) 220%

Hint: The Bernoulli equation and isentropic tables are useful.

### PROBLEM 3

A 24 in diameter pipeline carries sweet crude oil (specific gravity 0.86, kinematic viscosity $4 \times 10^{-5}$ ft²/sec) to a refinery at a rate of 14,000 gal/min. The flow inside the pipe is most nearly

    (A) laminar, Re = $2.0 \times 10^3$
    (B) turbulent, Re = $8.5 \times 10^3$
    (C) turbulent, Re = $41 \times 10^3$
    (D) turbulent, Re = $500 \times 10^3$

Hint: Do not waste time worrying whether the 24 in diameter is nominal or actual.

### PROBLEM 4

Water flows in a horizontal, square cast-iron conduit, 2 in on each side. The flow rate is 0.05 ft³/sec. The kinematic viscosity of water is $1.08 \times 10^{-5}$ ft²/sec, and the density of water is 62.4 lbm/ft³. Assume all minor losses are incorporated in a minor loss coefficient, $K$, equal to 0.8. The expected pressure drop over 40 ft of conduit length is most nearly

    (A) 0.14 psi
    (B) 0.17 psi
    (C) 0.19 psi
    (D) 0.30 psi

Hint: Use the extended Bernoulli equation.

### PROBLEM 5

Airflow velocity over a 1 ft by 1 ft horizontal flat plate is numerically determined to be

$$\mathrm{v}(x,y) = (5-x)\left(\frac{y}{5} + \frac{y^2}{10} + \frac{y^3}{100}\right)$$

$x$ and $y$ are in feet, and the function $\mathrm{v}(x,y)$ is in feet per second. $x$ is measured from the midpoint of the leading edge of the plate, and $y$ is measured up from the plate. Air pressure is 1 atm and temperature is 70°F.

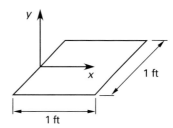

The total shear force acting on the upper surface of the plate is most nearly

    (A) $3.4 \times 10^{-7}$ lbf
    (B) $3.8 \times 10^{-7}$ lbf
    (C) $1.1 \times 10^{-5}$ lbf
    (D) $1.2 \times 10^{-5}$ lbf

Hint: Use Newton's law of viscosity.

## PROBLEM 6

A 150 hp reduced crude pump in a refinery pumps 1700 gpm. The brake horsepower needed to pump 2500 gpm is most nearly

(A) 150 hp
(B) 220 hp
(C) 320 hp
(D) 480 hp

Hint: The relationship between the brake horsepower of a pump and its flow rate is nonlinear.

## PROBLEM 7

A pipeline carries crude oil (specific gravity 0.86, kinematic viscosity $250 \times 10^{-6}$ ft$^2$/sec). A 250 hp pump at the pipeline pumping station generates 450 ft of head. The increase in pressure at the pump discharge is most nearly

(A) 170 psi
(B) 200 psi
(C) 2000 psi
(D) 24,000 psi

Hint: Equate feet of head to static pressure.

## PROBLEM 8

The feedwater piping system of a boiler is designed for an operating pressure of 2200 psi and a temperature of 280°F at the outlet. The friction pressure drop in the piping system under the design conditions is 33 psi. The design flow rate is $1.4 \times 10^6$ lbm/hr. The velocity pressure differentials and the static pressure drops are negligible; the lowest mass flow rate is 740,000 lbm/hr. The lowest pressure drop in the feedwater piping is most nearly

(A) 9.2 lbf/in$^2$
(B) 17 lbf/in$^2$
(C) 33 lbf/in$^2$
(D) 120 lbf/in$^2$

Hint: Assume the friction pressure drop is proportional to the square of the velocity.

## PROBLEM 9

A centrifugal pump at an oil pipeline pumping station pumps 70,000 barrels/day of crude oil (specific gravity 0.86, kinematic viscosity $250 \times 10^{-6}$ ft$^2$/sec) through a 12 in schedule-40 steel pipe to a booster pump inside a refinery 10 mi away. The pump develops 956 ft of head.

The motor torque supplied to the pump shaft running at 1750 rpm is 1500 lbf-ft. The pump's efficiency is most nearly

(A) 16%
(B) 67%
(C) 85%
(D) 98%

Hint: Calculate the water horsepower of the pump.

## PROBLEM 10

A water system with a capacity of 100 gpm has a total head of 240 ft at the suction line of a single-stage, two-pole, centrifugal pump. The maximum attainable efficiency for the pump is most nearly

(A) 29%
(B) 40%
(C) 52%
(D) 70%

Hint: Calculate the specific speed of the pump.

## PROBLEM 11

Water is being drawn through a siphon from a tank as shown. The temperature of the tank is 70°F, and the atmospheric pressure is 14.7 psia. The velocity of the water at the lower end of the siphon is 30 ft/sec.

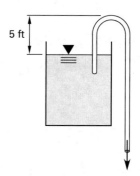

The pressure in the siphon at its highest point is most nearly

(A) 0.44 psia
(B) 6.5 psia
(C) 8.7 psia
(D) 13 psia

Hint: Choose a zero height reference that simplifies the Bernoulli equation.

## PROBLEM 12

A pump can supply 20 ft of head to a water flow of 5 lbm/sec. The water enters from a stagnant lake and exits as a free jet 15 ft above the lake, both locations at atmospheric pressure. The maximum velocity that the 5 lbm/sec flow can have for this single pump is most nearly

(A) 13 ft/sec
(B) 18 ft/sec
(C) 36 ft/sec
(D) 320 ft/sec

Hint: Use the extended Bernoulli equation, and find the *maximum* velocity.

## PROBLEM 13

In the piping system shown, 100 gal/min of water is pumped from one tank to another.

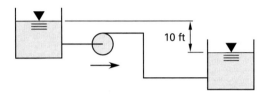

At this flow, the head loss due to piping and fittings is 90 ft. The pump has an efficiency of 60% and is driven by an electric motor with an efficiency of 90%. The fluid density is 50 lbm/ft$^3$. The power input to the motor is most nearly

(A) 0.38 hp
(B) 3.0 hp
(C) 3.4 hp
(D) 3.8 hp

Hint: Pump power depends on both pump head and flow rate.

## ENERGY/POWER SYSTEMS

## PROBLEM 14

A Carnot engine operates with a minimum temperature of 80°F, a maximum temperature of 1200°F, and a maximum pressure of 1300 psia. Assume that air behaves as an ideal gas. If 0.2 lbm of air is the working fluid and 5 Btu of heat is added to the air, the maximum cylinder volume is most nearly

(A) 0.10 ft$^3$
(B) 1.5 ft$^3$
(C) 2.0 ft$^3$
(D) 2.7 ft$^3$

Hint: Begin at the state where the pressure is a maximum.

## PROBLEM 15

A boiler generates 200,000 lbm/hr of 85% quality saturated steam at 1500 psia using 220°F feedwater. The energy absorbed by the steam is most nearly

(A) 180 MM Btu/hr
(B) 200 MM Btu/hr
(C) 220 MM Btu/hr
(D) 280 MM Btu/hr

Hint: Think about enthalpies.

## PROBLEM 16

An 800 gpm centrifugal pump generating 700 ft of head is pumping a light, hydrocarbon oil through a pipeline. If the pump is 50% efficient, the maximum possible temperature increase of the oil as it passes through the pump is most nearly

(A) 0.36°F
(B) 0.90°F
(C) 1.8°F
(D) 3.6°F

Hint: The specific heat, $c_p$, of the oil will be needed.

## PROBLEM 17

The number of independent quantities needed to specify the thermodynamic state of a system in equilibrium is equal to

(A) three
(B) the number of phases of the substance present in the system
(C) the number of possible phases a substance can have
(D) the number of possible work modes plus one

Hint: The number of independent quantities is not a fixed number.

## PROBLEM 18

Which of the following statements about the thermal efficiency of reversible heat engines are true?

I. The higher the temperature at which heat is added, the higher the efficiency.
II. The maximum thermal efficiency is achieved using a Carnot cycle.
III. The lower the temperature at which heat is rejected, the higher the efficiency.
IV. Different reversible engines operating between the same temperature limits have different thermal efficiencies.

(A) I and II
(B) I and IV
(C) II and III
(D) I, II, and III

Hint: This is a nonquantitative question. Start with the index of a thermodynamics book.

## PROBLEM 19

The $T$-$s$ diagram shown represents the air-standard power cycle of an internal combustion engine.

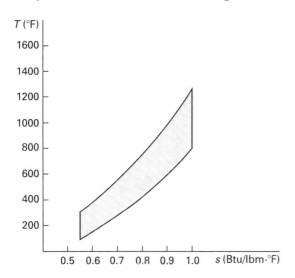

The thermal efficiency of this engine is most nearly

(A) 26%
(B) 36%
(C) 67%
(D) 92%

Hint: Internal combustion engines are evaluated using a particular power cycle.

## PROBLEM 20

An air compressor operates at steady state. Air enters through a 1.0 ft² opening at a velocity of 20 ft/sec, a temperature of 77°F, and a pressure of 14.7 psia, and exits at a velocity of 6 ft/sec, a temperature of 360°F, and a pressure of 100 psia. The power input to the compressor is 150 hp. Assuming ideal air properties, the heat input to the compressor is most nearly

(A) −48,000 Btu/hr
(B) −19,000 Btu/hr
(C) −18,000 Btu/hr
(D) 0 Btu/hr

Hint: Neglect potential energy changes.

# Depth Problems

## PRINCIPLES

### PROBLEM 21

An intermediate-depth underwater mine designed for coastal defense consists of a hollow, right-circular cylinder that encases an MK42 torpedo. Because of the tether design, the system can be modeled as a double pendulum as shown, with one pivot at the bottom of the case and one at the top of the anchor. The first two natural frequencies of oscillation for this particular system are 0.310 rad/sec and 1.81 rad/sec.

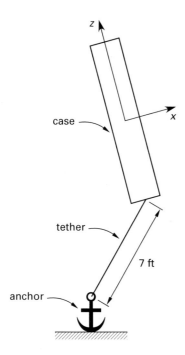

| case properties | |
|---|---|
| length | 122 in |
| outside diameter | 23 in |
| inside diameter | 22.25 in |
| weight | 1526 lbf |
| center of gravity (measured from bottom of case) | 71.2 in |
| $I_x$ | 728 ft-lbf-sec$^2$ |
| $I_y$ | 728 ft-lbf-sec$^2$ |
| $I_z$ | not available |

For seawater at 50°F the current velocities (in knots) that will cause the mine to oscillate at its two natural frequencies are most nearly

(A) 0.280 kt, 1.64 kt
(B) 0.473 kt, 2.76 kt
(C) 1.76 kt, 10.3 kt
(D) 3.36 kt, 19.6 kt

Hint: The Reynolds number will be needed, so it will be necessary to assume a reasonable number for the current velocity.

### PROBLEM 22

A preliminary design to transport 1000 gpm of water from a reservoir to an open tank is shown. The line includes four shutoff gate valves, one swing check valve, and two 45° elbows. If all the fittings are flange fittings, the pump's total dynamic head is most nearly

(A) 140 ft
(B) 150 ft
(C) 180 ft
(D) 210 ft

Hint: The friction head loss for each section of pipe must be calculated separately.

*Illustration for Problem 22*

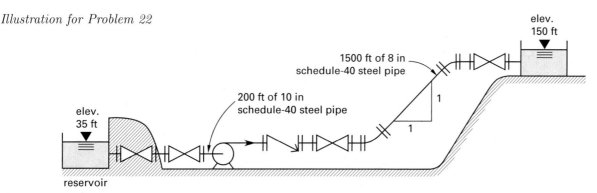

(not to scale)

## PROBLEM 23

In the oil supply system shown, oil is pumped from the reservoir to a bearing via the manifold. The oil exiting the bearing is then collected and funneled back into the reservoir.

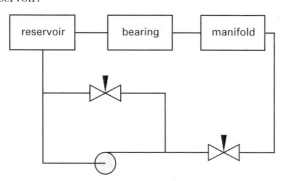

It is desired that the bearing be supplied with oil at 15 psi gauge. The primary functions of the two needle valves are to

I. control the pressure at the pump intake and the pressure at the manifold
II. control the flow rate to the manifold and the pressure at the manifold
III. control the flow rate to the pump intake and the flow rate to the manifold
IV. control the pressure at the pump intake and the flow rate to the manifold

(A) I only
(B) II only
(C) III only
(D) III and IV

Hint: This is a nonquantitative problem. Think about what happens if one needle valve is completely closed while the other is completely open.

## PROBLEM 24

A polyethylene downspout attachment that directs rainwater away from a house is shown. The polyethylene unrolls when it rains and then automatically re-rolls itself once the rain stops.

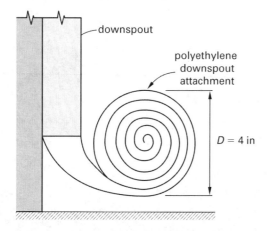

If the spring steel inside exerts a 0.5 in-lbf moment to re-roll the polyethylene, the height of water in a standard 3 in by 4 in downspout needed to initially begin unrolling the polyethylene is most nearly

(A) 0.29 in
(B) 0.58 in
(C) 1.2 in
(D) 6.9 in

Hint: This is a fluid statics problem.

## PROBLEM 25

The air-transfer pipe on an air compressor has an outside diameter of 0.5 in and is made of steel with a thermal conductivity of 27 Btu-ft/hr-ft$^2$-°F. Under typical operating conditions, the outer surface of the pipe is at a temperature of 150°F and is exposed to ambient air at 80°F, with a convection coefficient of 3 Btu/hr-ft$^2$-°F. Attached to the pipe are 100 equally spaced annular fins (also known as circular or transverse fins) 0.125 in thick and 1.25 in long.

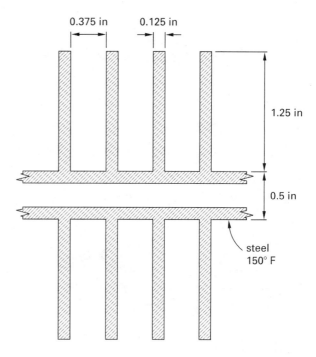

The heat transfer rate from all the fins is most nearly

(A) 870 Btu/hr
(B) 1300 Btu/hr
(C) 1700 Btu/hr
(D) 2200 Btu/hr

Hint: A fin efficiency graph will be helpful.

## PROBLEM 26

A 10 in schedule-80 pipe transports saturated water at an initial temperature of 400°F and a flow rate of 7230 lbm/hr. The inside film coefficient can be assumed to be infinite. The outside film coefficient can also be assumed to be infinite, and the outside temperature is 66°F. The pipe is insulated with calcium silicate so that the temperature loss is limited to 1°F per 40 ft of pipe. The minimum thickness of the insulation is most nearly

(A) 3.4 in
(B) 3.8 in
(C) 6.9 in
(D) 8.8 in

Hint: Compared with that of the insulation, the thermal resistance of the steel is negligible.

## PROBLEM 27

The air in one large room is at 70°F, 1.0 atm of pressure, and 80% relative humidity. A 50 ft long duct with a cross-sectional area of 2 ft$^2$ connects this room to another at 60°F, 1.0 atm, and 10% relative humidity. The diffusivity of water vapor in air is 1.0 ft$^2$/hr. What is the rate of vapor transfer by diffusion between the two rooms?

(A) $1.3 \times 10^{-5}$ lbm/hr
(B) $1.7 \times 10^{-5}$ lbm/hr
(C) $3.4 \times 10^{-5}$ lbm/hr
(D) $4.3 \times 10^{-5}$ lbm/hr

Hint: The pressure is uniform.

## PROBLEM 28

Air flows across a 4 ft$^2$ area covered with water. The air is at 70°F and 50% relative humidity, while the water is maintained at 60°F. The air velocity is such that the average mass transfer coefficient is 0.10 ft/sec. The rate of evaporation from the water surface is most nearly

(A) 0.091 lbm/hr
(B) 0.37 lbm/hr
(C) 0.47 lbm/hr
(D) 1.1 lbm/hr

Hint: Mass transfer by diffusion is proportional to the difference in vapor density.

## PROBLEM 29

A pump rated for 350 gpm and 200 ft of head is required for use as a refinery boiler feedwater pump that will run continuously year-round. The manufacturer's performance curves for a particular pump are shown. If the pump is oversized by 20%, the increased yearly electrical cost at $0.07/kW-hr for the oversized pump is most nearly

(A) $0
(B) $915
(C) $2750
(D) $5490

Hint: Be sure to increase both the flow rate and head by 20%, then treat the oversize requirement as a new rating and design accordingly.

*Illustration for Problem 29*

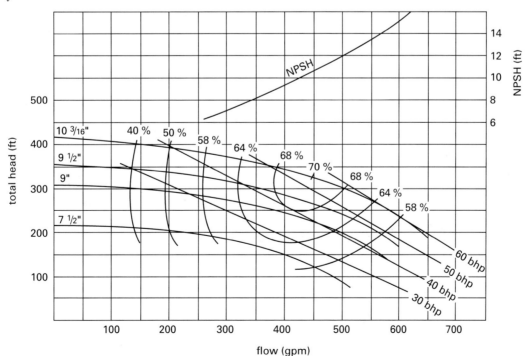

Reprinted/adapted with permission from Flowserve.

## PROBLEM 30

A company is considering the purchase of a piece of equipment costing $130,000. If purchased, the equipment will generate $45,000 of net profit for each of the next four years. This net profit represents all profits less operating and maintenance costs, but before taxes. At the end of these four years, the equipment will be salvaged for $10,000. Straight-line depreciation is used. The company's tax rate is 45% and the minimum attractive rate of return is 8%. What is the resulting present value, and should this purchase be made?

(A) −$5500; do not make purchase
(B) −$4000; do not make purchase
(C) $4000; make purchase
(D) $5500; make purchase

Hint: A high present value indicates a more attractive option.

## PROBLEM 31

6 in schedule-40 pipe is used to transport 50,000 lbm/hr of 550°F steam at 500 psia. The pressure drop per 100 ft of pipe is most nearly

(A) 2.2 lbf/in$^2$
(B) 3.6 lbf/in$^2$
(C) 6.1 lbf/in$^2$
(D) 26 lbf/in$^2$

Hint: A typical friction factor for turbulent flow in steel pipe is 0.02.

## PROBLEM 32

Air enters a converging-diverging nozzle with a stagnation temperature of 240°F and a stagnation pressure of 200 psia. In the diverging portion of the nozzle, a normal shock is located where M is equal to 1.8. If the exit area is three times larger than the throat area, what is the exit Mach number?

(A) 0.20
(B) 0.25
(C) 2.4
(D) 2.6

Hint: Isentropic flow and normal shock tables are useful.

## PROBLEM 33

A jet of water from a stationary nozzle impinges on a moving block and vane of mass 50 lbm with a turning angle of 30°. The vane is initially at rest against the ground with a coefficient of static friction of 0.1. The water exits the nozzle and hits the vane at 150 ft/sec through an area of 0.005 ft$^2$.

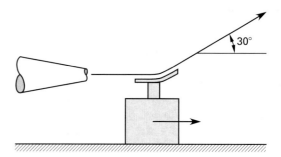

The initial acceleration of the mass is most nearly

(A) 0.27 ft/sec$^2$
(B) 8.6 ft/sec$^2$
(C) 19 ft/sec$^2$
(D) 29 ft/sec$^2$

Hint: Neglect friction between the vane and fluid.

## PROBLEM 34

Consider two Brayton cycles. In both systems, there is isentropic compression in the compressor (A-B) and constant-pressure heat addition in the combustor (B-C).

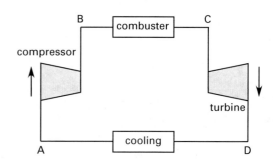

In system 1, an isentropic expansion occurs in the turbine (C-D), followed by constant-pressure cooling (D-A). In system 2, the turbine isentropic efficiency is 75%. The two systems operate such that points A, B, and C are the same on a $T$-$s$ diagram for each. Which of the following statements are true?

I. The thermal efficiency of system 1 is higher than that of system 2.
II. The work of the compressor is the same in systems 1 and 2.
III. More heat is rejected during cooling in system 1 than in system 2.
IV. The work out of the turbine is greater in system 1 than in system 2.

(A) I and II only
(B) III and IV only
(C) I, II, and IV only
(D) I, II, III, and IV

Hint: No calculations are required.

## PROBLEM 35

If the steam generated in a process has a circulation ratio of five, which of the following statements is true?

(A) The steam-water mixture contains five times more water than steam.
(B) The steam-water mixture contains five times more steam than water.
(C) 1 lbm of the steam-water mixture contains 0.2 lbm of water.
(D) 1 lbm of the steam-water mixture contains 0.2 lbm of steam.

Hint: Think about steam quality.

## PROBLEM 36

A combustion furnace uses 2 lbmol of $H_2$ and 1 lbmol of $O_2$ at 537°R and produces 2 lbmol of water vapor in the exhaust. If the products have an average specific heat of 1.15 Btu/lbm-°R, the maximum possible exit temperature is most nearly

(A) 1200°R
(B) 1800°R
(C) 5000°R
(D) 5500°R

Hint: The maximum temperature is equal to the adiabatic flame temperature.

## PROBLEM 37

An unknown amount of $C_3H_8$ (propane) is burned in the air. An Orsat (dry product) analysis gives results of 6.6% $O_2$, 7.7% $CO_2$, and 2.2% CO. The excess air in the reaction is most nearly

(A) 17%
(B) 25%
(C) 33%
(D) 90%

Hint: Determine the stoichiometric reaction.

## PROBLEM 38

An 18 in diameter pipeline carries crude oil (sp. gr. 0.94) up from the Gulf Coast to a refinery at a rate of 8000 gpm. Inside the refinery a 180° horizontal bend occurs along a horizontal stretch of the pipeline. This bend has a volume of 100 gal and is held in place by 32 1 in UNC bolts, that is, 16 bolts in each flange at the two ends of the bend. If the bend weighs 627 lbf, the shear stress in each bolt is most nearly

(A) 25 psi
(B) 31 psi
(C) 56 psi
(D) 110 psi

Hint: Pay attention to the flow direction.

## PROBLEM 39

The maximum horsepower that can be transmitted by a 2.5 in diameter, solid steel cooling tower fan shaft running at 200 rpm is most nearly

(A) 80 hp
(B) 150 hp
(C) 190 hp
(D) 290 hp

Hint: Use a failure criterion.

## PROBLEM 40

The valve on a fuel gas pipe has been accidentally broken off its nipple. As a temporary fix, a cylindrical plug has been inserted into the nipple, and a steel band has been strapped around the plug and the pipe and tightened so that the initial tension in the band is 100 lbf, as shown.

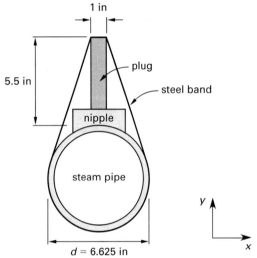

If the steel band has a breaking strength of 300 lbf, the maximum pressure that can be exerted on the plug by the fuel gas inside the pipe before the band breaks is most nearly

(A) 230 psi
(B) 360 psi
(C) 490 psi
(D) 730 psi

Hint: Use a free-body diagram.

## PROBLEM 41

Mulch is being transferred from one conveyor belt to another as shown, at a rate of 77 tons/hr. The speed of the upper conveyor belt is 400 ft/min, and the speed of the lower conveyor belt is 850 ft/min.

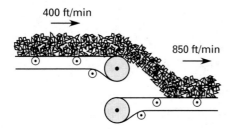

The coefficient of friction between the belts and the rollers is 0.25, and the tension in the bottom belt of the lower conveyor system is 100 lbf. The tension in the upper belt of the lower conveyor system to the left of where the mulch lands is most nearly

(A) 56 lbf
(B) 120 lbf
(C) 190 lbf
(D) 210 lbf

Hint: Use 180° as the contact angle between the belt and the roller.

## PROBLEM 42

Part of the repair to a center pipe-to-flange connection in a reactor is shown. The main exit pipe (not shown) was placed over the short piece of pipe labeled in the drawing. These two pipes were then welded together to complete the repair. The connection is held by 38 half-inch UNF 20 bolts. The bolts are ASTM A354 grade BC. The construction is such that the bolts are subjected to a purely axial load and are designed to fail before the weld.

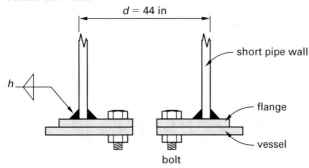

If the weld is to be twice as strong as the bolts and if the allowable shear stress in the weld is 35 ksi, the minimum weld depth is most nearly

(A) 0.250 in
(B) 0.500 in
(C) 0.625 in
(D) 0.750 in

Hint: Pay attention to how the external loading in a fillet weld is carried.

## PROBLEM 43

A steel pipe buried in the earth carries natural gas (weight density 0.046 lbf/ft$^3$) over a highway as shown.

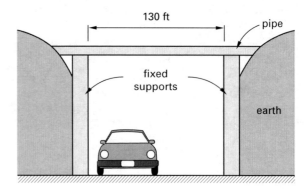

If 30 in schedule-30 pipe is used, the maximum bending stress in the pipe is most nearly

(A) 4.0 ksi
(B) 7.9 ksi
(C) 12 ksi
(D) 48 ksi

Hint: Pay particular attention to the ends of the pipe.

## PROBLEM 44

Air at a dry-bulb temperature of 84°F has a relative humidity of 50%. The heat needed to heat 2500 lbm/hr of this air to 102°F without changing the moisture content is most nearly

(A) $6.5 \times 10^3$ Btu/hr
(B) $11 \times 10^3$ Btu/hr
(C) $23 \times 10^3$ Btu/hr
(D) $38 \times 10^3$ Btu/hr

Hint: A psychrometric chart will be useful.

## PROBLEM 45

The block diagram for the control of an electrically actuated hydraulic valve is shown.

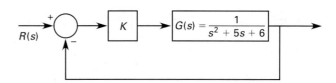

The largest value of the gain, $K$, that will maintain a stable, closed-loop system is most nearly

(A) 0.25
(B) 2.5
(C) 6.0
(D) $\infty$

Hint: The open-loop transfer function is important.

## PROBLEM 46

The turbine shown has an output of 10 MW and is 90% efficient. Saturated steam at 1000 psia flows from the boiler through the turbine and exhausts to a condenser at 2 psia. The condensate is subcooled 6°F in the condenser. The condensate flow is 100,000 lbm/hr.

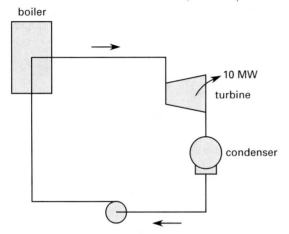

The thermal efficiency of the cycle is most nearly

(A) 9.1%
(B) 28%
(C) 31%
(D) 34%

Hint: Sufficient accuracy can be obtained by neglecting pump work.

## EQUIPMENT
### PROBLEM 47

A refinery boiler feedwater pump operating at 3550 rpm has a 9.5 in impeller and performance curves as shown.

If the pump needs to achieve a rating of 550 gpm and 350 ft, the speed at which it should be run is most nearly

(A) 3070 rpm
(B) 4150 rpm
(C) 4740 rpm
(D) 5140 rpm

Hint: Use trial and error.

*Illustration for Problem 47*

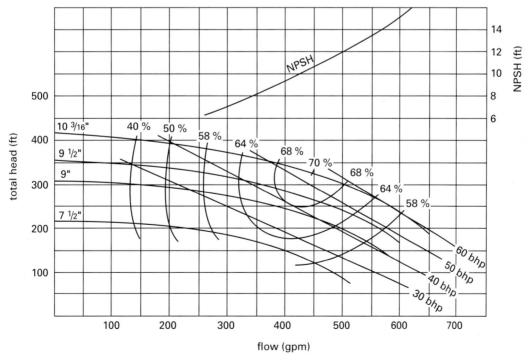

Reprinted/adapted with permission from Flowserve.

## PROBLEM 48

A crude overhead product pump operating at 3550 rpm has the characteristics shown.

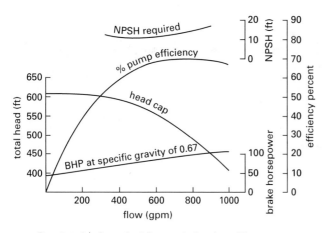

Reprinted/adapted with permission from Flowserve.

It is connected to a system requiring 200 ft of head when handling 500 gpm of crude oil. The flow rate in the system when the pump is running at 3550 rpm is most nearly

(A) 220 gpm
(B) 300 gpm
(C) 800 gpm
(D) 1000 gpm

Hint: Draw the system curve.

## PROBLEM 49

A pump pumps water from a lake to two water distribution systems. The feed points for both systems are located 95 ft above the lake. Pressure drops due to friction in the pipes are estimated at 23.2 psi in the pump suction line and 67.5 psi in the pump discharge line up to the T junction. From the T junction to feed point A, the pressure drop is 43.8 psi, and from the T junction to feed point B, the pressure drop is 28.5 psi. The shutoff valves each have a pressure drop of 5 psi.

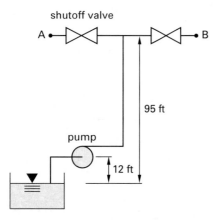

Point A has a volumetric flow requirement of 860 gal/min, and point B has a requirement of 350 gal/min. The energy losses due to pressure differential between the suction and the discharge lines are negligible. The velocity head between suction and discharge is also negligible. The pump has a mechanical efficiency of 95%. The pump brake horsepower required to supply the two systems with the required volumetric flow is most nearly

(A) 31.2 hp
(B) 102 hp
(C) 105 hp
(D) 113 hp

Hint: Use the sum of inverses to combine the pressure drops in parallel lines.

## PROBLEM 50

An impeller with a 2.0 ft diameter operates at its best efficiency point. At this efficiency point, its speed is 440 rpm and it produces a flow rate of 200 gpm, a head of 180 ft, and 17.0 hp of power. If the impeller is clipped to 1.5 ft and still operates at 440 rpm, the new flow rate, head, and power will be most nearly

(A) 150 gpm, 140 ft, and 13 hp
(B) 200 gpm, 140 ft, and 9.6 hp
(C) 84 gpm, 76 ft, and 7.2 hp
(D) 150 gpm, 100 ft, and 7.2 hp

Hint: The impeller is clipped; it is not geometrically similar.

## PROBLEM 51

140 gpm of water at 180°F is pumped out of a vented tank. For this pump at this flow, the manufacturer requires 5.0 ft of net positive suction head (NPSH).

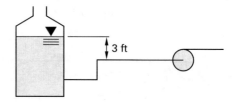

The head loss between the tank and the pump suction is given by the equation

$$h_{f(s)} = 1.50 \times 10^{-5} L Q^{1.852}$$

$h_{f(s)}$ is in feet, $L$ is the pipeline length in feet, and $Q$ is the flow rate in gallons per minute. Surroundings are at 1.0 atm. The maximum permissible pipeline length, if cavitation is to be avoided, is most nearly

(A) 85 ft
(B) 100 ft
(C) 140 ft
(D) 230 ft

Hint: Compare required NPSH and available NPSH.

## PROBLEM 52

A manufacturer of electrically actuated four-way hydraulic valves uses the following equation to compute the flow through its valves.

$$Q_R = K I_m \sqrt{P_v}$$

$Q_R$ is the rated flow in gallons per minute, $K$ is a valve constant, $I_m$ is the maximum DC current supplied to the valve (15 milliamps), and $P_v$ is the pressure drop across the valve. The manufacturer rates its valves assuming a 1000 psi drop across the valve, and has 1, 5, 10, and 15 gpm valves available, cost increasing with size. In a particular application, the supply and peak load pressures are 1500 psi and 1100 psi respectively. The least expensive valve that will supply a 23 in$^3$/sec flow under these conditions is the

(A) 1 gpm valve
(B) 5 gpm valve
(C) 10 gpm valve
(D) 15 gpm valve

Hint: The valve constant is key here.

## PROBLEM 53

Which two of the following are valid pipe fitting classifications?

I. NPT
II. SAE
III. UNC
IV. ASME

(A) I and II
(B) I and III
(C) II and III
(D) III and IV

Hint: Do not confuse screw threads with pipe threads.

## PROBLEM 54

The throttling area in a needle valve is a function of the

I. seat opening diameter
II. needle lift
III. cone height
IV. cone half angle

(A) I and IV
(B) II and III
(C) I, II, and IV
(D) I, III, and IV

Hint: The throttling areas for a needle valve and a poppet valve are calculated using the same equation.

## PROBLEM 55

An orifice must control the flow rate of hydraulic fluid at 130°F to 1.5 gpm ± 5% with a pressure differential of 1000 psi. The fluid specific gravity is 1.09 at the specified temperature. The required flow resistance is most nearly

(A) 400 lohms
(B) 910 lohms
(C) 3000 lohms
(D) 12,000 lohms

Hint: The lohm, or liquid ohm, is defined as the restriction that permits a flow of 100 gpm of water at 80°F with a pressure drop of 25 psi.

## PROBLEM 56

A steam turbine generating system that operates at an overall efficiency of 75% with a back pressure of 300 psia is required to generate 9 MW. Superheated steam at 1300 psia and 940°F is used. The amount of steam required is most nearly

(A) $66.7 \times 10^3$ lbm/hr
(B) $113 \times 10^3$ lbm/hr
(C) $171 \times 10^3$ lbm/hr
(D) $228 \times 10^3$ lbm/hr

Hint: A turbine can be modeled as an isentropic process.

## PROBLEM 57

Which of the following can be done to increase the net positive suction head available (NPSHA) to a pump?

I. Use a larger diameter pipe.
II. Place a throttling valve in the intake line.
III. Pressurize the supply tank.
IV. Increase the height of the supply tank.

(A) I and IV
(B) II and IV
(C) I, II, and IV
(D) I, III, and IV

Hint: This is a nonquantitative question. Start with the index of a fluids book.

## PROBLEM 58

A 44 in long, 1.25 in diameter compressor shaft is positioned in vee blocks. A dial indicator is placed on the shaft to determine the shaft runout. The readings of the indicator are given in the following table.

| angle (deg) | reading (0.1 mil) | angle (deg) | reading (0.1 mil) |
|---|---|---|---|
| 0   | 0.0 | 190 | 1.0 |
| 10  | 0.0 | 200 | 0.5 |
| 20  | 0.0 | 210 | 0.5 |
| 30  | 0.5 | 220 | 0.5 |
| 40  | 0.5 | 230 | 0.0 |
| 50  | 0.5 | 240 | 0.0 |
| 60  | 0.5 | 250 | 0.0 |
| 70  | 1.0 | 260 | −0.5 |
| 80  | 1.0 | 270 | −0.5 |
| 90  | 1.0 | 280 | −0.5 |
| 100 | 1.0 | 290 | −0.5 |
| 110 | 1.0 | 300 | −1.0 |
| 120 | 1.5 | 310 | −1.0 |
| 130 | 1.5 | 320 | −0.5 |
| 140 | 1.5 | 330 | −0.5 |
| 150 | 1.5 | 340 | −0.5 |
| 160 | 1.5 | 350 | 0.0 |
| 170 | 1.0 | 360 | 0.0 |
| 180 | 1.0 | | |

The mechanical runout of the shaft is most nearly

(A) 0.036 mil
(B) 0.10 mil
(C) 0.15 mil
(D) 0.25 mil

Hint: The mechanical runout is a difference.

## PROBLEM 59

A three-phase motor operating at 93% efficiency and pulling 100 A is driving a high-pressure boiler feedwater pump in a refinery that pumps 500 gpm with an 850 psi differential. If the motor voltage is 2300 V and the power factor is 0.87, the pump efficiency is most nearly

(A) 57%
(B) 77%
(C) 86%
(D) 99%

Hint: When efficiencies are given, begin looking at power.

## PROBLEM 60

A fluid leaving the impeller of a 24 in diameter centrifugal compressor must have a tangential velocity of 300 ft/sec to be properly compressed. The mass flow rate is 900 lbm/sec and the operating speed of the compressor is 6300 rpm. The horsepower of the compressor is most nearly

(A) 170 hp
(B) 1600 hp
(C) 5700 hp
(D) 10,000 hp

Hint: Assume a single-stage compressor to get an estimate of the power.

## PROBLEM 61

The flow characteristics of a typical centrifugal compressor are shown.

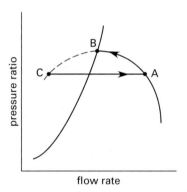

The cycle ABCA is the phenomenon known as

(A) surging
(B) stalling
(C) starving
(D) throttling

Hint: Look carefully at the flow rate during this cycle.

## PROBLEM 62

A pump running at 3550 rpm must supply 340 ft of head. To run the pump at 75% efficiency, the flow rate must be most nearly

(A) 200 gpm
(B) 500 gpm
(C) 1000 gpm
(D) greater than is possible given the operating conditions

Hint: The specific speed of the pump will be useful.

## PROBLEM 63

An axial flow pump running at sea level is used to transfer water at 80°F from an open-air tank to a higher level. The friction loss in the suction pipe is 5 ft. The net positive suction head available is 10 ft. The suction lift is most nearly

(A) 10 ft
(B) 18 ft
(C) 26 ft
(D) 30 ft

Hint: Be sure to use units of "feet absolute."

## PROBLEM 64

The evaporator section of a heat recovery steam generator is made up of a drum boiler that generates the steam, a tube bank heat exchanger, and pipes that carry the water and steam. Preheated water arrives in the drum at 600°F. The drum operates at 1900 psia and produces 95% quality steam at a rate of 540,000 lbm/hr. Each tube in the tube bank has a useful heat-exchange surface of 5.3 ft$^2$ and an overall coefficient of heat transfer of 218.2 Btu/hr-ft$^2$-°F. The average temperature difference between the water in the tube bank and flue gas is 32°F. Because of heat losses, only 92% of the heat transmitted to the drum contributes to the generation of steam. The number of tubes needed to maintain the required steam production is most nearly

(A) 6700 tubes
(B) 8000 tubes
(C) 8400 tubes
(D) 260,000 tubes

Hint: Find the heat transfer rate needed to produce the steam.

## PROBLEM 65

A fan rated at 4700 standard cubic feet per minute (SCFM) is placed in 70°F air at an altitude of 6000 ft. The actual airflow rate of the fan at this temperature and altitude is most nearly

(A) 3800 ft$^3$/min
(B) 5900 ft$^3$/min
(C) 6100 ft$^3$/min
(D) 9200 ft$^3$/min

Hint: A density correction factor for altitude will be needed.

## PROBLEM 66

A compressor is needed to compress natural gas from 35 psig to 200 psig, at a temperature of 60°F and a rate of $1.7 \times 10^6$ standard cubic feet per 24 hours. The compressor will be located 3000 ft above sea level, and the ratio of specific heat of the gas is 1.28. The horsepower of the compressor must be most nearly

(A) 95 hp
(B) 120 hp
(C) 160 hp
(D) 200 hp

Hint: The compression ratio is important.

## PROBLEM 67

A noncooled main air blower at a refinery compresses 50,000 cfm of atmospheric air to 53 psia. The compressor efficiency is 79%. The motor shaft horsepower required to operate the main air blower is most nearly

(A) 110 hp
(B) 1800 hp
(C) 4000 hp
(D) 6300 hp

Hint: Think adiabatic power.

## PROBLEM 68

A cooling tower design requires a water area of 608 ft$^2$. Since this size is not available, a two-cell tower having a water area of 576 ft$^2$ (2 times 12 ft by 24 ft) will be used. The increase in the air rate needed to handle the increase in the water load for this two-cell tower is most nearly

(A) 2.1%
(B) 5.2%
(C) 6.0%
(D) 15%

Hint: A correlation equation based on experimental data will be useful.

## PROBLEM 69

A boiler horsepower is

I. a unit of power equal to approximately 33,475 Btu/hr, or 9.81 kW
II. equal to one mechanical horsepower
III. the boiler capacity needed to produce one unit of horsepower using a turbine operating at 98% efficiency
IV. the amount of power needed to convert 34.5 lbm/hr of feedwater at 212°F and atmospheric pressure to dry, saturated steam at the same temperature and pressure

(A) I only
(B) I and III
(C) I and IV
(D) II and III

Hint: Boiler horsepower is traditionally used to rate and purchase packaged fire tube boilers.

## SYSTEMS

### PROBLEM 70

The smallest nominal schedule-40 pipe diameter that should be used to transport 1750 gpm of water through a city's water supply system is most nearly

(A) 4.0 in
(B) 12 in
(C) 32 in
(D) 48 in

Hint: A rule of thumb will provide a quick estimate.

### PROBLEM 71

A 70,000 lbm steel waste-heat boiler contains 45,000 lbm of water at a temperature of 90°F. The gases at 1300°F enter the unit at a rate of 100,000 lbm/hr. The boiler surface area is 30,000 ft$^2$. The flue gas specific heat is 0.25 Btu/lbm-°F. The overall heat transfer coefficient is 5 Btu/ft$^2$-hr-°F. The minimum time required to bring the boiler to 212°F is most nearly

(A) 12 min
(B) 14 min
(C) 23 min
(D) 31 min

Hint: This is similar to a heat exchanger problem.

### PROBLEM 72

A centrifugal pump is to deliver 300 gpm of hot water from an elevated tank. The water in the tank is saturated liquid at 120 psig and is delivered to the pump through 12 in schedule-40 pipe. The pressure at the pump inlet is 200 psig. The critical cavitation number is 1.2. The flow velocity inside the pump at which cavitation will begin is most nearly

(A) 75 ft/sec
(B) 110 ft/sec
(C) 170 ft/sec
(D) 220 ft/sec

Hint: The vapor pressure will be needed.

### PROBLEM 73

A run-around heat recovery system uses a coolant loop and two heat exchangers to extract energy from warm air exhausted from a building and transfer the energy to cold air brought into the building.

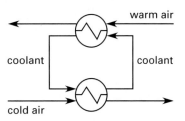

Before passing through the system, the warm air is at 75°F and the cold air is at 10°F. Air flows through the system in both directions at 36,000 lbm/hr, and cooling flows at 15,000 lbm/hr. Heat is removed from the outgoing air at 200,000 Btu/hr.

The overall heat transfer coefficient for each heat exchanger is 50 Btu/hr-ft$^2$-°F. The heat exchangers are a crossflow type with both fluids unmixed. The specific heats are 0.24 Btu/lbm-°F for the air and 1.0 Btu/lbm-°F for the coolant. Heat loss from the coolant line and energy added by the coolant pump can be neglected.

The coolant temperatures in the left and right halves, respectively, of the loop are most nearly

(A) 75°F and 10°F
(B) 49°F and 36°F
(C) 54°F and 30°F
(D) 30°F and 54°F

Hint: Due to the symmetry of this arrangement, the average of the coolant temperatures on the two sides of the loop is equal to the average of the temperatures of the incoming warm and cold air.

### PROBLEM 74

The quality of 1800 psia saturated steam required to absorb 800 Btu/lbm from a boiler when the feedwater temperature is 280°F is most nearly

(A) 20%
(B) 30%
(C) 80%
(D) 90%

Hint: Do not forget about the energy in the feedwater.

### PROBLEM 75

The supplementary firing system of a heat recovery steam generator burns natural gas in the flue gas with a thermal efficiency of 55%. The higher heating value of the natural gas used is 17,000 Btu/lbm. The mass flow rate of the flue gas through the firing system is 3,180,000 lbm/hr. The combustion process in the firing system occurs at constant pressure, and the specific heat of the flue gas before and after combustions is approximately 0.248 Btu/lbm-°F. The natural gas is fed into the firing system at a temperature of 60°F and a mass flow rate of 65,000 lbm/hr. The specific heat of the natural gas is 0.692 Btu/lbm-°F. If the flue gas enters the firing system at a temperature of 980°F, its temperature as it leaves is most nearly

(A) 700°F
(B) 1700°F
(C) 1800°F
(D) 2100°F

Hint: Start by finding the thermal energy provided by the combustion of the natural gas.

## PROBLEM 76

Air flowing at a rate of 1 lbm/sec at 70°F and 14.7 psia is to be compressed to 200 psia. Which of the following statements are true?

I. An isentropic process would require the least amount of power.
II. An adiabatic process must also be an isentropic process.
III. The amount of power required, within 10%, for an isentropic process would be 200 hp.

(A) I only
(B) II only
(C) III only
(D) I and III only

Hint: A typical isothermal process uses intercooling between staged compressors.

## PROBLEM 77

The cooling water requirement for the lube oil system of an 1100 hp gas engine with a water-cooled exhaust manifold is most nearly

(A) 17.6 gpm
(B) 132 gpm
(C) 484 gpm
(D) 1320 gpm

Hint: Use an approximation here.

## PROBLEM 78

A closed feedwater heater subsystem is shown. Refrigerant-12, condensed to a saturated liquid at 0°F and with a quality of zero, is pumped isentropically into the heater at a rate of 1 lbm/sec and a pressure of 100 psia.

Once heated, the R-12 leaves the heater, still as a saturated liquid with a quality of zero and still at a pressure of 100 psia. The energy to heat it comes from R-12 returning in the form of a superheated gas, flowing at a rate of 0.5 lbm/sec, a temperature of 100°F, and a pressure of 120 psia.

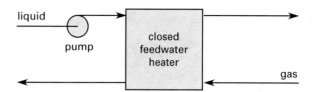

Assume that all the energy lost by the flow of gas is gained by the flow of liquid, that there are no pressure drops in the feedwater heater, and that potential and kinetic energies can be neglected. The quality of the superheated gas as it leaves the feedwater heater is most nearly

(A) 27%
(B) 40%
(C) 84%
(D) 86%

Hint: Use the energy balance to calculate the enthalpy of the gas leaving the heater.

## CODES AND STANDARDS
## PROBLEM 79

A full journal bearing with a length/diameter ($L/d$) ratio of 1, a diameter of 1.5 in, and a radial clearance of 2 mils supports a journal carrying a 495 lbf radial load running at 3600 rpm. If the inlet temperature of the SAE 30 supply oil is 120°F, the temperature rise of the oil is most nearly

(A) 18°F
(B) 26°F
(C) 36°F
(D) 56°F

Hint: Design charts from Raimondi and Boyd will be useful.

## PROBLEM 80

A section of 3 in, seamless, schedule-40, SA-106 grade B pipe in a crude distillation unit operates at 250 psi and 600°F. The pipe was installed in 1990, and subsequent ultrasonic thickness readings are given in the following table.

| year | thickness (in) |
|---|---|
| 1995 | 0.208 |
| 1998 | 0.200 |
| current year | 0.175 |

The remaining corrosion allowance is most nearly

(A) 0.12 in
(B) 0.15 in
(C) 0.16 in
(D) 0.19 in

Hint: Use ASME code B31.3 and/or API 570.

## PROBLEM 81

A nozzle of SA-106 grade B material with a 4.5 in inner diameter abuts a seamless section of a vessel. The vessel has a maximum allowable working pressure of 250 psi and a design temperature of 500°F. The vessel's inside diameter is 68 in and its walls are 0.625 in thick. If the nozzle provides 0.518 in$^2$ of reinforcement, the amount of additional reinforcement needed is most nearly

(A) 1.0 in$^2$
(B) 1.2 in$^2$
(C) 2.7 in$^2$
(D) 3.4 in$^2$

Hint: Refer to ASME Sec. VIII.

## PROBLEM 82

A boiler produces 200,000 lbm/hr of saturated steam at 1400 psig set pressure. The nozzle area of the steam relief valve is most nearly

(A) 2.7 in$^2$
(B) 3.1 in$^2$
(C) 8.9 in$^2$
(D) 24 in$^2$

Hint: This problem describes a safety valve.

## PROBLEM 83

A centrifugal feedwater pump has its seals cooled using piping plan 12 to inject water at the seal interface. The heat generation at the seal faces is 1.5 kW. The minimum allowable injection flow rate is most nearly

(A) 0.92 L/min
(B) 1.8 L/min
(C) 3.7 L/min
(D) 7.7 L/min

Hint: Use the American Petroleum Institute standard API 682.

## PROBLEM 84

An unstayed, circular blind flange is needed to isolate a vessel with an internal design pressure specified at 200 psi. The inside diameter of the vessel is 2 ft and the allowable stress is 13,000 psi. If the attachment factor is 0.17, the required thickness of the blind flange is most nearly

(A) 0.10 in
(B) 1.2 in
(C) 3.0 in
(D) 6.6 in

Hint: Refer to ASME Sec. VIII, Div. I.

## PROBLEM 85

A flat, 3 ft diameter, unstayed head is to be designed for a life of 30 years and a design pressure of 150 psi. The maximum allowable stress is 17,500 psi, and the attachment factor is 0.20. If the corrosion rate is 8 mil/yr and the joint efficiency is 1, the design thickness of the head is most nearly

(A) 0.24 in
(B) 0.36 in
(C) 1.7 in
(D) 3.6 in

Hint: Refer to ASME Sec. VIII, Div. I.

# Breadth Solutions

## HYDRAULICS AND FLUIDS

### SOLUTION 1

The Mach number of an object is the ratio of the object's speed to the speed of sound in the medium through which it is flying. This definition makes it necessary to determine the speed of sound, $a$, in air at 50,000 ft. This can be done using the formula

$$a = \sqrt{kg_c RT}$$

Since air is treated as an ideal gas, the ratio of specific heats, $k$, and the specific gas constant, $R$, are both constant. For air,

$$k = 1.4$$
$$R = 53.35 \text{ ft-lbf/lbm-°R}$$

The given temperature at 50,000 ft is

$$T = 393°R$$

The speed of sound at 50,000 ft is

$$a = \sqrt{kRTg_c}$$
$$= \sqrt{(1.4)\left(32.2 \ \frac{\text{ft-lbm}}{\text{lbf-sec}^2}\right)\left(53.35 \ \frac{\text{ft-lbf}}{\text{lbm-°R}}\right)(393°R)}$$
$$= 972 \text{ ft/sec}$$

The Mach number is

$$M = \frac{v}{a} = \frac{\left(1200 \ \frac{\text{mi}}{\text{hr}}\right)\left(\frac{1 \text{ hr}}{3600 \text{ sec}}\right)\left(5280 \ \frac{\text{ft}}{\text{mi}}\right)}{972 \ \frac{\text{ft}}{\text{sec}}}$$
$$= 1.81 \quad (1.8)$$

**The answer is (C).**

### Why Other Options Are Wrong

(A) This incorrect answer results when the Mach number ratio is inverted.

(B) This incorrect answer results when the speed of sound at sea level (1130 ft/sec) is used to compute the Mach number.

(D) This incorrect answer results when the ratio of specific heats is neglected in calculating the speed of sound.

### SOLUTION 2

At sea level (0 ft altitude) international standard atmosphere conditions are 518.7°R and 14.7 psia.

For compressible flow at Mach 1, the ratio of static to stagnation pressure, $p/p_0$, is 0.5283.

Thus, for compressible flow the maximum stagnation pressure is

$$p_{2,c} = \frac{p}{0.5283} = \frac{14.7 \ \frac{\text{lbf}}{\text{in}^2}}{0.5283} = 27.8 \text{ psi}$$

For incompressible flow, Bernoulli's equation is needed.

$$\frac{p_1}{\rho} + \frac{v_1^2}{2g_c} + \frac{z_1 g}{g_c} = \frac{p_{2,i}}{\rho} + \frac{v_2^2}{2g_c} + \frac{z_2 g}{g_c}$$

For a streamline connecting a point away from the plane (1) to the stagnation point at the nose (2), $z_1$ is equal to $z_2$, and $v_2$ is equal to 0. Solving for $p_{2,i}$,

$$p_{2,i} = p_1 + \frac{\rho v_1^2}{2g_c}$$

$p_1$ is 14.7 psia. For air, the gas constant, $R$, is 53.35 ft-lbf/lbm-°R. Density can be calculated from the equation

$$\rho = \frac{p}{RT} = \frac{\left(14.7 \ \frac{\text{lbf}}{\text{in}^2}\right)\left(12 \ \frac{\text{in}}{\text{ft}}\right)^2}{\left(53.35 \ \frac{\text{ft-lbf}}{\text{lbm-°R}}\right)(518.7°R)}$$
$$= 0.076 \text{ lbm/ft}^3$$

To find $v_1$, only the speed of sound is needed, because the plane is flying at Mach 1.

$$v_1 = a = \sqrt{kg_c RT}$$

$$= \sqrt{(1.4)\left(32.2 \frac{\text{ft-lbm}}{\text{lbf-sec}^2}\right) \times \left(53.35 \frac{\text{ft-lbf}}{\text{lbm-°R}}\right)(518.7°\text{R})}$$

$$= 1116.9 \text{ ft/sec}$$

Then,

$$p_{2,i} = p_1 + \frac{\rho v_1^2}{2g_c}$$

$$= 14.7 \frac{\text{lbf}}{\text{in}^2} + \frac{\left(0.076 \frac{\text{lbm}}{\text{ft}^3}\right)\left(1116.9 \frac{\text{ft}}{\text{sec}}\right)^2}{(2)\left(32.2 \frac{\text{ft-lbm}}{\text{lbf-sec}^2}\right)}$$

$$\times \left(\frac{1 \text{ ft}}{12 \text{ in}}\right)^2$$

$$= 24.9 \text{ psi}$$

The fractional error is

$$\varepsilon = \frac{|p_{2,c} - p_{2,i}|}{p_{2,c}}$$

$$= \frac{\left|27.8 \frac{\text{lbf}}{\text{in}^2} - 24.9 \frac{\text{lbf}}{\text{in}^2}\right|}{27.8 \frac{\text{lbf}}{\text{in}^2}}$$

$$= 0.104 \quad (10\%)$$

**The answer is (A).**

**Why Other Options Are Wrong**

(B) This incorrect answer occurs when $k$ is omitted in calculating $v_1$.

(C) This incorrect answer occurs when $\rho v_1^2$ is divided by $g_c$ instead of $2g_c$.

(D) This incorrect answer occurs when $p$ is multiplied by 0.5283 instead of divided by it.

## SOLUTION 3

The size of the Reynolds number determines whether the flow is laminar or turbulent.

$$\text{Re} = \frac{vd}{\nu}$$

The average velocity of the flow inside the pipe is found from the formula

$$Q = vA$$

$$v = \frac{Q}{A} = \frac{\left(14{,}000 \frac{\text{gal}}{\text{min}}\right)\left(\frac{1 \text{ ft}^3}{7.48 \text{ gal}}\right)\left(\frac{1 \text{ min}}{60 \text{ sec}}\right)}{\left(\frac{\pi}{4}\right)(24 \text{ in})^2 \left(\frac{1 \text{ ft}}{12 \text{ in}}\right)^2}$$

$$= 9.9 \text{ ft/sec}$$

The Reynolds number is

$$\text{Re} = \frac{\left(9.9 \frac{\text{ft}}{\text{sec}}\right)(24 \text{ in})\left(\frac{1 \text{ ft}}{12 \text{ in}}\right)}{4 \times 10^{-5} \frac{\text{ft}^2}{\text{sec}}} = 500 \times 10^3$$

**The answer is (D).**

**Why Other Options Are Wrong**

(A) This incorrect answer results when a flow rate of 1400 gal/min is mistakenly used, $\pi d$ is used for the area, and the inches-to-feet conversion is neglected when the flow velocity is calculated.

(B) This incorrect answer results when the number of gallons per minute is divided by 3600 instead of 60 when calculating v.

(C) This incorrect answer results when the inches-to-feet conversion is neglected.

## SOLUTION 4

Let point 1 be 40 ft upstream of point 2. The extended Bernoulli equation is

$$\frac{p_1}{\rho} + \frac{v_1^2}{2g_c} + \frac{z_1 g}{g_c} + E_A = \frac{p_2}{\rho} + \frac{v_2^2}{2g_c} + \frac{z_2 g}{g_c} + E_E + E_f + E_m$$

For a horizontal pipe,

$$z_1 = z_2$$

The pipe is continuous, so

$$v_1 = v_2$$

There are no pumps or turbines, so

$$E_A = E_E = 0$$

Solving the extended Bernoulli equation for the drop in pressure, then, gives

$$p_1 - p_2 = \rho(E_f + E_m)$$

$E_m$ represents minor losses and can be calculated from the minor loss coefficient, $K$.

$$E_m = h_m\left(\frac{g}{g_c}\right) = Kh_v\left(\frac{g}{g_c}\right) = K\left(\frac{v^2}{2g}\right)\left(\frac{g}{g_c}\right)$$
$$= K\left(\frac{v^2}{2g_c}\right)$$

$E_f$ represents frictional losses and can be evaluated from the equation

$$E_f = h_f\left(\frac{g}{g_c}\right) = \left(\frac{fLv^2}{2dg}\right)\left(\frac{g}{g_c}\right) = \frac{fLv^2}{2dg_c}$$

The average velocity in the pipe, v, is

$$v = \frac{Q}{A} = \frac{0.05\ \frac{ft^3}{sec}}{(2\ in)^2 \left(\frac{1\ ft}{12\ in}\right)^2} = 1.80\ ft/sec$$

Thus,

$$E_m = K\left(\frac{v^2}{2g_c}\right) = (0.8)\left(\frac{\left(1.80\ \frac{ft}{sec}\right)^2}{(2)\left(32.2\ \frac{ft\text{-}lbm}{lbf\text{-}sec^2}\right)}\right)$$
$$= 4.03 \times 10^{-2}\ ft\text{-}lbf/lbm$$

For a noncircular pipe, the hydraulic diameter, $d_h$, must be used. For a square duct with side length $l$,

$$d_h = \frac{4A}{P} = \frac{4l^2}{4l} = l = 2\ in$$

To find the friction factor, $f$, the Reynolds number of the flow, Re, must first be found, using the hydraulic diameter.

$$Re = \frac{vd}{\nu} = \frac{vd_h}{\nu} = \frac{\left(1.80\ \frac{ft}{sec}\right)(2\ in)\left(\frac{1\ ft}{12\ in}\right)}{1.08 \times 10^{-5}\ \frac{ft^2}{sec}}$$
$$= 2.78 \times 10^4$$

For a cast-iron pipe, the specific roughness, $\varepsilon$, is $8.0 \times 10^{-4}$ ft. The relative roughness is then

$$\frac{\varepsilon}{d_h} = \frac{8.0 \times 10^{-4}\ ft}{(2\ in)\left(\frac{1\ ft}{12\ in}\right)} = 4.8 \times 10^{-3}$$

For these values, a Moody friction factor chart gives a friction factor of 0.033. The frictional loss term can now be calculated.

$$E_f = \frac{fLv^2}{2dg_c} = \frac{fLv^2}{2d_hg_c}$$
$$= \frac{(0.033)(40\ ft)\left(1.80\ \frac{ft}{sec}\right)^2}{(2)(2\ in)\left(\frac{1\ ft}{12\ in}\right)\left(32.2\ \frac{ft\text{-}lbm}{lbf\text{-}sec^2}\right)}$$
$$= 3.99 \times 10^{-1}\ lbf\text{-}ft/lbm$$

The pressure drop is

$$p_1 - p_2 = \rho(E_f + E_m)$$
$$= \left(62.4\ \frac{lbm}{ft^3}\right)\left(\begin{array}{l}3.99 \times 10^{-1}\ \frac{ft\text{-}lbf}{lbm} \\ + 4.03 \times 10^{-2}\ \frac{ft\text{-}lbf}{lbm}\end{array}\right)$$
$$\times \left(\frac{1\ ft}{12\ in}\right)^2$$
$$= 1.90 \times 10^{-1}\ lbf/in^2 \quad (0.19\ psi)$$

**The answer is (C).**

Why Other Options Are Wrong

(A) This incorrect answer results when the relative roughness is calculated without converting the pipe's hydraulic diameter to feet.

(B) This incorrect answer results when minor losses are neglected.

(D) This incorrect answer results when the area of the pipe is calculated as a circular cross section of diameter 2 in, the hydraulic diameter.

**SOLUTION 5**

Newton's law of viscosity applied at the surface (where $y$ equals 0) is

$$\tau_{y=0} = \mu\frac{dv}{dy} = \mu\left(\frac{\partial v(x,y)}{\partial y}\right)_{y=0}$$
$$= \mu(5-x)\left(\frac{1}{5} + \frac{y}{5} + \frac{3y^2}{100}\right)_{y=0}$$
$$= \mu(5-x)\left(\frac{1}{5}\ sec^{-1}\right)$$

Air at 70°F has a viscosity ($\mu$) of $3.80 \times 10^{-7}$ lbf-sec/ft$^2$. $v(x,y)$ has units of ft/sec, so the derivative of $v(x,y)$ with respect to $y$ has units of sec$^{-1}$, and $\tau$ has units of lbf/ft$^2$. After integrating, the unit area in square feet can be multiplied into the result, giving an answer in lbf.

The shear force is obtained by integrating the shear stress, $\tau$, over the area of the plate, noting that the differential area, $dA$, is equal to $dxdz$ and noting that $x$ goes from 0 ft to 1 ft and $z$ goes from $-0.5$ ft to 0.5 ft.

$$F = \int \tau_{y=0} dA$$

$$= \int_{z=-0.5 \text{ ft}}^{0.5 \text{ ft}} \int_{x=0 \text{ ft}}^{1 \text{ ft}} \frac{\mu}{5}(5-x)dx\,dz$$

$$= \int_{-0.5 \text{ ft}}^{0.5 \text{ ft}} dz \int_{0 \text{ ft}}^{1 \text{ ft}} \frac{\mu}{5}(5-x)dx$$

$$= z\bigg|_{-0.5 \text{ ft}}^{0.5 \text{ ft}} \left(\frac{\mu}{5}\left(5x - \frac{x^2}{2}\right)\right)\bigg|_{0 \text{ ft}}^{1 \text{ ft}}$$

$$= 0.9\mu \text{ ft}^2/\text{sec}$$

$$= \left(0.9 \frac{\text{ft}^2}{\text{sec}}\right)\left(3.80 \times 10^{-7} \frac{\text{lbf-sec}}{\text{ft}^2}\right)$$

$$= 3.42 \times 10^{-7} \text{ lbf} \quad (3.4 \times 10^{-7} \text{ lbf})$$

**The answer is (A).**

Why Other Options Are Wrong

(B) This incorrect answer results when the shear stress is evaluated for $x = 0$ and $y = 0$ and the result is multiplied by the area of the plate (1 ft$^2$).

(C) This answer results when an incorrect viscosity of $1.22 \times 10^{-5}$ lbm/ft-sec is used.

(D) This answer results when the shear stress is evaluated for $x = 0$ and $y = 0$ and an incorrect viscosity of $1.22 \times 10^{-5}$ lbm/ft-sec is used.

## SOLUTION 6

This problem is solved using the pump affinity laws.

$$\frac{Q_1}{Q_2} = \frac{n_1}{n_2}$$

$$\frac{\text{BHP}_1}{\text{BHP}_2} = \left(\frac{n_1}{n_2}\right)^3$$

Substitute the flow ratio for the speed ratio in the brake horsepower equation, substitute the given values, and solve for the new brake horsepower, BHP$_2$.

$$\frac{\text{BHP}_1}{\text{BHP}_2} = \left(\frac{Q_1}{Q_2}\right)^3$$

$$\frac{150 \text{ hp}}{\text{BHP}_2} = \left(\frac{1700 \text{ gpm}}{2500 \text{ gpm}}\right)^3$$

$$\text{BHP}_2 = 477 \text{ hp} \quad (480 \text{ hp})$$

**The answer is (D).**

Why Other Options Are Wrong

(A) This answer is incorrect. The brake horsepower of the pump is a function of the pump's flow rate. Therefore, an increase in the flow rate requires an increase in the brake horsepower.

(B) This incorrect answer results when the flow rate ratio is not cubed in the calculation of the brake horsepower.

(C) This incorrect answer results when the flow rate ratio is squared instead of cubed in the calculation of the brake horsepower.

## SOLUTION 7

A static column of liquid of height $h$ (feet of head) is equivalent to a certain pressure. The relationship between the two is given by

$$p = h\gamma(\text{SG})$$

The pressure at the discharge for a head of 450 ft of crude oil is

$$p = (450 \text{ ft})\left(62.4 \frac{\text{lbf}}{\text{ft}^3}\right)\left(\frac{1 \text{ ft}}{12 \text{ in}}\right)^2 (0.86)$$

$$= 168 \text{ lbf/in}^2 \quad (170 \text{ psi})$$

Note that head and pressure are not interchangeable terms since head is potential energy and pressure is force per unit area.

**The answer is (A).**

Why Other Options Are Wrong

(B) This incorrect answer results when the specific gravity of crude oil is not included in the calculation of $\gamma$. Therefore, this is the pressure if the liquid being pumped were water.

(C) This incorrect answer results when 1 ft$^2$ is converted to 12 in$^2$ instead of $(12 \text{ in})^2$.

(D) This incorrect answer results when pressure is calculated in pounds per square foot, not pounds per square inch.

## SOLUTION 8

The friction head at the design point is

$$h_{f,d} = \frac{fLv_d^2}{2dg} = K v_d^2 = K\left(\frac{\dot{m}_d}{\rho A}\right)^2$$

The friction head at the off-design point is

$$h_{f,\text{off}} = \frac{fLv_{\text{off}}^2}{2dg} = K\left(\frac{\dot{m}_{\text{off}}}{\rho A}\right)^2$$

Solving these two equations for the pressure head at the off-design conditions gives

$$h_{f,\text{off}} = h_{f,d} \frac{\dot{m}_{\text{off}}^2}{\dot{m}_d^2}$$

The pressure drop at the off-design is

$$\Delta p_{\text{off}} = \Delta p_d \frac{\dot{m}_{\text{off}}^2}{\dot{m}_d^2}$$

$$= \left(33 \; \frac{\text{lbf}}{\text{in}^2}\right) \frac{\left(740{,}000 \; \frac{\text{lbm}}{\text{hr}}\right)^2}{\left(1.4 \times 10^6 \; \frac{\text{lbm}}{\text{hr}}\right)^2}$$

$$= 9.22 \; \text{psi} \quad (9.2 \; \text{psi})$$

**The answer is (A).**

### Why Other Options Are Wrong

(B) This incorrect answer results when the exponent on the mass flow rate is neglected.

(C) This incorrect answer results when the pressure drop for all operating conditions is incorrectly assumed to be equal to the design pressure drop, and therefore independent of the mass flow rate.

(D) This incorrect answer results when the proportionality equation is inverted as

$$\Delta p_{\text{off}} = \Delta p_d \frac{\dot{m}_d^2}{\dot{m}_{\text{off}}^2}$$

## SOLUTION 9

The pump's efficiency is the ratio of the pump's water horsepower (WHP) to the pump's brake horsepower (BHP).

$$\eta = \frac{\text{WHP}}{\text{BHP}}$$

The brake horsepower can be found using the pump's shaft torque and speed.

$$\text{BHP} = T\omega$$

$$= (1500 \; \text{ft-lbf}) \left(1750 \; \frac{\text{rev}}{\text{min}}\right) \left(2\pi \; \frac{\text{rad}}{\text{rev}}\right) \left(\frac{1 \; \text{min}}{60 \; \text{sec}}\right)$$

$$\times \left(\frac{1 \; \text{hp}}{550 \; \frac{\text{ft-lbf}}{\text{sec}}}\right)$$

$$= 500 \; \text{hp}$$

The water horsepower can be found using the formula

$$\text{WHP} = Q\gamma h_A$$

The flow rate is

$$Q = \left(70{,}000 \; \frac{\text{barrels}}{\text{day}}\right) \left(42 \; \frac{\text{gal}}{\text{barrel}}\right) \left(\frac{1 \; \text{ft}^3}{7.48 \; \text{gal}}\right)$$

$$\times \left(\frac{1 \; \text{day}}{86{,}400 \; \text{sec}}\right)$$

$$= 4.549 \; \text{ft}^3/\text{sec}$$

The water horsepower is

$$\text{WHP} = Q\gamma(\text{SG})h_A$$

$$= \left(4.549 \; \frac{\text{ft}^3}{\text{sec}}\right) \left(62.4 \; \frac{\text{lbf}}{\text{ft}^3}\right) (0.86)(956 \; \text{ft})$$

$$\times \left(\frac{1 \; \text{hp}}{550 \; \frac{\text{ft-lbf}}{\text{sec}}}\right)$$

$$= 424 \; \text{hp}$$

The efficiency of the pump is

$$\eta = \frac{424 \; \text{hp}}{500 \; \text{hp}} = 0.848 \quad (85\%)$$

**The answer is (C).**

### Why Other Options Are Wrong

(A) This incorrect answer results when the 70,000 barrels per day is converted to gallons per second and then the reciprocal of the efficiency is used.

(B) This incorrect answer results when the specific gravity of crude oil is not included in the WHP calculation. Therefore, the water horsepower is calculated for water instead of crude oil.

(D) This incorrect answer results when 0.68 is used for the specific gravity instead of 0.86.

## SOLUTION 10

The pump's efficiency is found using a graph that gives the pump efficiency as a function of specific speed and capacity. The specific speed is a dimensionless design parameter used to describe and classify pump impellers. The specific speed is found using the formula

$$n_s = \frac{n\sqrt{Q}}{h^{0.75}}$$

A dimensional analysis of the specific speed equation will show that $n_s$ is not truly dimensionless. Converting $Q$ to cubic feet per minute and multiplying $h$ by the acceleration of gravity, $g$, before raising it to the 0.75 power will make the equation dimensionless. The convention in the centrifugal pump industry, however, is not to include these constant terms.

The speed of the 60 Hz, two-pole motor needed to drive the pump can be found using the equation

$$n = \frac{7200 \text{ rpm}}{N} = \frac{7200 \text{ rpm}}{2} = 3600 \text{ rpm}$$

The true operating speed of a two-pole motor ranges from 3450 to 3550 rpm. The reduced speed is due to slippage between the rotor and stator.

Using 3550 rpm as the speed, the specific speed for this pump is

$$n_s = \frac{(3550 \text{ rpm})\sqrt{100 \frac{\text{gal}}{\text{min}}}}{(240 \text{ ft})^{0.75}} = 582$$

Using a "typical" pump efficiency graph, the pump's maximum attainable efficiency is found to be 52%. Since pump efficiency data is manufacturer-specific, variations in this result can be expected.

**The answer is (C).**

Why Other Options Are Wrong

(A) This incorrect answer results when the 100 gpm is converted to cubic feet per minute.

(B) This incorrect answer results when the speed of a four-pole motor, 1780 rpm, is used instead the speed of a two-pole motor.

(D) This incorrect answer results when the efficiency is read from the 1000 gpm curve instead of the 100 gpm curve.

## SOLUTION 11

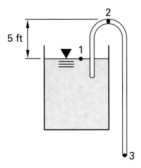

The total energy is constant, so the Bernoulli equation is

$$\frac{p_1}{\rho} + \frac{v_1^2}{2g_c} + \frac{z_1 g}{g_c} = \frac{p_2}{\rho} + \frac{v_2^2}{2g_c} + \frac{z_2 g}{g_c}$$

From the law of conservation of mass for an incompressible fluid, the velocity at point 2 is the same as the velocity at point 3. Assuming the velocity at point 1 is 0 ft/sec, and using point 1 as the zero height reference,

$$\frac{p_1}{\rho} + 0 + 0 = \frac{p_2}{\rho} + \frac{v_2^2}{2g_c} + \frac{z_2 g}{g_c}$$

Solving for the pressure at point 2 yields

$$p_2 = p_1 - \left(\frac{v_2^2}{2g_c} + \frac{z_2 g}{g_c}\right)\rho$$

The density of water at 70°F is 62.3 lbm/ft³.

$$p_2 = 14.7 \frac{\text{lbf}}{\text{in}^2}$$
$$- \left(\frac{\left(30 \frac{\text{ft}}{\text{sec}}\right)^2}{(2)\left(32.2 \frac{\text{ft-lbm}}{\text{lbf-sec}^2}\right)} + \frac{(5 \text{ ft})\left(32.2 \frac{\text{ft}}{\text{sec}^2}\right)}{32.2 \frac{\text{ft-lbm}}{\text{lbf-sec}^2}}\right)$$
$$\times \left(62.3 \frac{\text{lbm}}{\text{ft}^3}\right)\left(\frac{1 \text{ ft}}{12 \text{ in}}\right)^2$$
$$= 6.5 \text{ lbf/in}^2 \quad (6.5 \text{ psia})$$

**The answer is (B).**

Why Other Options Are Wrong

(A) This incorrect solution results from dividing by $g_c$ instead of $2g_c$ in the kinetic energy term.

(C) This incorrect solution results from omitting the potential energy term $zg/g_c$.

(D) This incorrect solution results from omitting the kinetic energy term $v^2/2g_c$.

## SOLUTION 12

Define location 1 at the stagnant lake surface and location 2 at the free-jet exit 15 ft higher. The extended Bernoulli equation is

$$\frac{p_1}{\rho} + \frac{v_1^2}{2g_c} + \frac{z_1 g}{g_c} + E_A = \frac{p_2}{\rho} + \frac{v_2^2}{2g_c} + \frac{z_2 g}{g_c} + E_E + E_f + E_m$$

At the stagnant lake surface the velocity, $v_1$, is zero. Let the height at the lake surface, $z_1$, be zero; then the height at the exit, $z_2$, is 15 ft. The pressure at the lake surface is atmospheric and the same value as at the free jet's exit, so the pressure terms cancel.

Using these values, the equation becomes

$$E_A = \frac{v_2^2}{2g_c} + \frac{(15 \text{ ft})g}{g_c} + E_E + E_f + E_m$$

Solving for $v_2$,

$$v_2 = \sqrt{2g_c\left(E_A - \frac{(15 \text{ ft})g}{g_c} - E_E - E_f - E_m\right)}$$

No energy is extracted in this system (by a turbine, for example), so $E_E$ is zero. $E_f$ and $E_m$ represent losses; to maximize the velocity, these are set to zero. $E_A$ represents energy added by the pump and can be calculated from the pump head, $h_A$.

$$E_A = \frac{gh_A}{g_c} = \frac{\left(32.2 \frac{\text{ft}}{\text{sec}^2}\right)(20 \text{ ft})}{32.2 \frac{\text{ft-lbm}}{\text{lbf-sec}^2}} = 20 \text{ ft-lbf/lbm}$$

The maximum exit velocity can now be calculated.

$$v_{2,\text{max}} = \sqrt{(2)\left(32.2 \frac{\text{ft-lbm}}{\text{lbf-sec}^2}\right) \times \left(20 \frac{\text{ft-lbf}}{\text{lbm}} - \frac{(15 \text{ ft})\left(32.2 \frac{\text{ft}}{\text{sec}^2}\right)}{32.2 \frac{\text{ft-lbm}}{\text{lbf-sec}^2}}\right)}$$

$$= 17.9 \text{ ft/sec} \quad (18 \text{ ft/sec})$$

**The answer is (B).**

Why Other Options Are Wrong

(A) This incorrect answer results when the velocities in the Bernoulli equation are divided by $g_c$ instead of $2g_c$.

(C) This incorrect answer results when the 15 ft elevation is omitted.

(D) This incorrect answer results when the square root in the equation for $v_2$ is omitted.

## SOLUTION 13

The head the pump must provide is simply the friction loss minus the elevation change, because the liquid level in the discharge tank is lower than that of the supply tank (apply Bernoulli equation). So, the pump head is

$$h_A = h_f - \Delta z = 90 \text{ ft} - 10 \text{ ft}$$
$$= 80 \text{ ft}$$

The flow converted to cubic feet per minute is

$$Q = \left(100 \frac{\text{gal}}{\text{min}}\right)\left(\frac{1 \text{ ft}^3}{7.48 \text{ gal}}\right) = 13.4 \text{ ft}^3/\text{min}$$

The required power is

$$P = \left(\frac{\gamma h_A Q}{\eta_{\text{pump}} \eta_{\text{motor}}}\right)\left(\frac{g}{g_c}\right)$$

$$= \left(\frac{\left(50 \frac{\text{lbm}}{\text{ft}^3}\right)(80 \text{ ft})\left(13.4 \frac{\text{ft}^3}{\text{min}}\right)}{(0.60)(0.90)}\right)$$

$$\times \left(\frac{32.2 \frac{\text{ft}}{\text{sec}^2}}{32.2 \frac{\text{ft-lbm}}{\text{lbf-sec}^2}}\right)\left(\frac{1 \text{ hp-min}}{33{,}000 \text{ ft-lbf}}\right)$$

$$= 3.0 \text{ hp}$$

**The answer is (B).**

Why Other Options Are Wrong

(A) This incorrect answer results when friction loss is neglected.

(C) This incorrect answer results when the elevation change is neglected.

(D) This incorrect answer results when the elevation change is added to the friction loss instead of being subtracted.

## ENERGY/POWER SYSTEMS

### SOLUTION 14

The $p$-$V$ diagram for a Carnot cycle is shown.

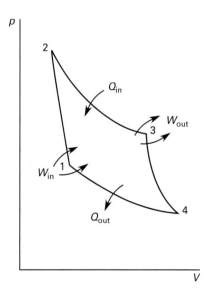

The maximum pressure and temperature occur at state 2, and the minimum pressure and temperature occur at state 4. Therefore, state 2 is where the minimum cylinder volume occurs, and state 4 is where the maximum cylinder volume occurs. The ideal gas law can be used to find the volume at state 2. Then the cycle process can be followed from state 2 to state 3, and then to state 4 where the maximum cylinder volume can be found.

At state 2,

$$p_2 V_2 = m R T_2$$

$$\left(1300 \ \frac{\text{lbf}}{\text{in}^2}\right) \left(12 \ \frac{\text{in}}{\text{ft}}\right)^2 V_2 = (0.2 \text{ lbm}) \left(53.35 \ \frac{\text{ft-lbf}}{\text{lbm-°R}}\right)$$
$$\times (1200°\text{F} + 460°)$$

$$V_2 = 0.095 \text{ ft}^3$$

The heat added to the air is known, so the volume at state 3 can be found using the equation

$$q_{in} = p_2 V_2 \ln \frac{V_3}{V_2}$$

$$(5 \text{ Btu}) \left(778 \ \frac{\text{ft-lbf}}{\text{Btu}}\right) = \left(1300 \ \frac{\text{lbf}}{\text{in}^2}\right) \left(12 \ \frac{\text{in}}{\text{ft}}\right)^2$$
$$\times (0.095 \text{ ft}^3) \ln \frac{V_3}{0.095 \text{ ft}^3}$$

$$V_3 = 0.12 \text{ ft}^3$$

The process from state 3 to state 4 is reversible and adiabatic, so $V_4$ can be found using the equation

$$\frac{V_4}{V_3} = \left(\frac{T_3}{T_4}\right)^{\frac{1}{k-1}}$$

The ratio of specific heats, $k$, for air is equal to 1.4. Thus, $V_4$, the maximum cylinder volume, is

$$\frac{V_4}{0.12 \text{ ft}^3} = \left(\frac{1200°\text{F} + 460°}{80°\text{F} + 460°}\right)^{\frac{1}{1.4-1}}$$

$$V_4 = 2.0 \text{ ft}^3$$

**The answer is (C).**

Why Other Options Are Wrong

(A) This incorrect answer results when the volume at state 2 is thought to be the maximum cylinder volume.

(B) This incorrect answer results when the conversion to Rankine is forgotten when calculating $V_2$.

(D) This incorrect answer results when the volume at state 2 is calculated using the universal gas constant instead of the specific gas constant and this volume is thought to be the maximum cylinder volume.

### SOLUTION 15

The energy absorbed by the steam can be found using

$$Q = \dot{m} \Delta h$$

The enthalpy of the feedwater at 220°F is

$$h_{\text{feed}} = 188.22 \text{ Btu/lbm}$$

The enthalpy of the 85% quality, 1500 psia saturated steam is calculated using the equation

$$h_{\text{steam}} = h_f + x h_{fg}$$

From the steam tables,

$$h_f = 611.5 \text{ Btu/lbm}$$
$$h_{fg} = 557.2 \text{ Btu/lbm}$$

The enthalpy of the steam is

$$h_{\text{steam}} = 611.5 \ \frac{\text{Btu}}{\text{lbm}} + (0.85) \left(557.2 \ \frac{\text{Btu}}{\text{lbm}}\right)$$
$$= 1085 \text{ Btu/lbm}$$

The energy absorbed by the steam is

$$Q = \left(200{,}000 \ \frac{\text{lbm}}{\text{hr}}\right) \left(1085 \ \frac{\text{Btu}}{\text{lbm}} - 188.22 \ \frac{\text{Btu}}{\text{lbm}}\right)$$
$$= 179 \times 10^6 \text{ Btu/hr} \quad (180 \text{ MM Btu/hr})$$

**The answer is (A).**

Why Other Options Are Wrong

(B) This incorrect answer results when the enthalpy of 120°F feedwater is used instead of the enthalpy of 220°F feedwater.

(C) This incorrect answer results when the initial feedwater enthalpy is not subtracted from the final steam enthalpy.

(D) This incorrect answer results when $h_g$ is used instead of $h_{fg}$.

### SOLUTION 16

This problem can be solved by equating the brake horsepower of the pump to the heat flow rate. The brake horsepower of the pump is

$$\text{BHP} = \frac{Q \gamma h}{\eta}$$

But $\dot{m} = Q \gamma$, so the brake horsepower equation can be rewritten as

$$\text{BHP} = \frac{\dot{m} h}{\eta}$$

From heat transfer, the heat flow rate is

$$Q = \dot{m} c_p \Delta T$$

The maximum possible temperature increase will occur when all the brake horsepower goes into heating the fluid. Converting Btu to ft-lbf and equating the heat flow equation to the brake horsepower equation gives

$$\dot{m} c_p \Delta T \left( 778 \, \frac{\text{ft-lbf}}{\text{Btu}} \right) = \frac{\dot{m} h}{\eta}$$

The specific heat for light, hydrocarbon oil is 0.5 Btu/lbm-°F. Because the specific heat is per pound of mass, the right side of the previous equation must be rewritten

$$\dot{m} c_p \Delta T \left( 778 \, \frac{\text{ft-lbf}}{\text{Btu}} \right) = \left( \frac{\dot{m} h}{\eta} \right) \left( \frac{g}{g_c} \right)$$

Solving for the temperature rise gives

$$\Delta T = \frac{hg}{\left( 778 \, \frac{\text{ft-lbf}}{\text{Btu}} \right) c_p \eta g_c}$$

$$= \frac{(700 \text{ ft}) \left( 32.2 \, \frac{\text{ft}}{\text{sec}^2} \right)}{\left( 778 \, \frac{\text{ft-lbf}}{\text{Btu}} \right) \left( 0.5 \, \frac{\text{Btu}}{\text{lbm-°F}} \right) (0.5) \left( 32.2 \, \frac{\text{ft-lbm}}{\text{lbf-sec}^2} \right)}$$

$$= 3.6 \degree \text{F}$$

**The answer is (D).**

Why Other Options Are Wrong

(A) This incorrect answer results when a head of 70 ft is used instead of 700 ft.

(B) This incorrect answer results when, in the calculation of the brake horsepower, the water horsepower is multiplied by the efficiency of the pump instead of divided by it.

(C) This incorrect answer results when the efficiency of the pump is neglected.

## SOLUTION 17

To specify the state of a substance, an intensive property such as pressure, electrical potential, or magnetic potential is needed for each work mode, in addition to an intensive property to describe heat interactions.

**The answer is (D).**

Why Other Options Are Wrong

(A) This answer is incorrect because the number of independent quantities needed varies depending on the substance in the system.

(B) This answer is incorrect because the number of phases of the substance is irrelevant.

(C) This answer is incorrect because the number of phases of the substance is irrelevant.

## SOLUTION 18

The thermal efficiency of a reversible heat engine operating between two reservoirs and exchanging only heat with these reservoirs is given by

$$\eta_{\text{th}} = 1 - \frac{T_L}{T_H}$$

$T_L$ and $T_H$ are the low and high temperatures, respectively. From this it can be seen that both statements I and III are true.

The above thermal efficiency represents the theoretical best efficiency that can be achieved. It can be shown that the Carnot cycle achieves this efficiency, so statement II is also true.

Statement IV is false because it can be shown that different reversible engines operating between the same temperature limits will have the same thermal efficiencies.

**The answer is (D).**

Why Other Options Are Wrong

(A) This is incorrect because statement III is also true.

(B) This is incorrect because statements II and III are also true and statement IV is false.

(C) This is incorrect because statement I is also true.

## SOLUTION 19

Internal combustion engines are evaluated on the basis of an air-standard Otto cycle. The processes involved are

1–2 isentropic compression
2–3 heat addition at constant volume
3–4 isentropic expansion
4–1 heat rejection at constant volume

The thermal efficiency of an Otto cycle can be calculated either from $T_1$ and $T_2$ or from $T_4$ and $T_3$.

$$\eta_{\text{th}} = 1 - \frac{T_1}{T_2} = 1 - \frac{T_4}{T_3}$$

The temperatures must be in degrees Rankine.

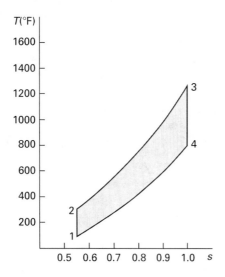

From the T-s diagram, $T_1 = 100°F$ and $T_2 = 300°F$. The efficiency is

$$\eta_{th} = 1 - \frac{T_1}{T_2} = 1 - \frac{100°F + 460°}{300°F + 460°} = 1 - \frac{560°R}{760°R}$$
$$= 0.26 \quad (26\%)$$

Alternatively,

$$\eta_{th} = 1 - \frac{T_4}{T_3} = 1 - \frac{800°F + 460°}{1250°F + 460°} = 1 - \frac{1260°R}{1710°R}$$
$$= 0.26 \quad (26\%)$$

**The answer is (A).**

Why Other Options Are Wrong

(B) This incorrect answer results when $T_3$ and $T_4$ are used but not converted to Rankine.

(C) This incorrect answer results when $T_1$ and $T_2$ are used but not converted to Rankine.

(D) This incorrect answer results when the Carnot efficiency is calculated and the temperatures are not converted to Rankine.

## SOLUTION 20

For a steady-state process, the first law of thermodynamics gives

$$Q - W + \dot{m}\left(h_1 + \frac{v_1^2}{2g_c} + gz_1\right) - \dot{m}\left(h_2 + \frac{v_2^2}{2g_c} + gz_2\right) = 0$$

Ignoring potential energy changes and solving for the heat added to the compressor,

$$Q = W - \dot{m}\left(h_1 - h_2 + \frac{v_1^2}{2g_c} - \frac{v_2^2}{2g_c}\right)$$

The mass flow rate for the incoming air is

$$\dot{m}_1 = \frac{A_1 v_1}{v_1} = \frac{A_1 v_1 p_1}{R T_1}$$

$$= \frac{(1.0 \text{ ft}^2)\left(20 \frac{\text{ft}}{\text{sec}}\right)\left(14.7 \frac{\text{lbf}}{\text{in}^2}\right)}{\left(53.35 \frac{\text{ft-lbf}}{\text{lbm-°R}}\right)(537°R)}$$

$$\times \left(3600 \frac{\text{sec}}{\text{hr}}\right)\left(12 \frac{\text{in}}{\text{ft}}\right)^2$$

$$= 5320 \text{ lbm/hr}$$

From property tables for low-pressure air, the enthalpy of the air at the inlet and outlet of the compressor, respectively, are

$$h_1 = 128.34 \text{ Btu/lbm}$$
$$h_2 = 196.69 \text{ Btu/lbm}$$

Substituting into the equation for the heat added to the compressor,

$$Q = (-150 \text{ hp})\left(2545 \frac{\text{Btu}}{\text{hp-hr}}\right) - \left(5320 \frac{\text{lbm}}{\text{hr}}\right)$$
$$\times \left( \begin{array}{c} 128.34 \frac{\text{Btu}}{\text{lbm}} - 196.69 \frac{\text{Btu}}{\text{lbm}} \\ + \frac{\left(20 \frac{\text{ft}}{\text{sec}}\right)^2 - \left(6 \frac{\text{ft}}{\text{sec}}\right)^2}{(2)\left(32.2 \frac{\text{ft-lbm}}{\text{lbf-sec}^2}\right)} \\ \times \left(\frac{1 \text{ Btu}}{778 \text{ ft-lbf}}\right) \end{array} \right)$$

$$= -18{,}167 \text{ Btu/hr} \quad (-18{,}000 \text{ Btu/hr})$$

**The answer is (C).**

Why Other Options Are Wrong

(A) This incorrect solution results from omitting the conversion factor from ft-lbf to Btu.

(B) This incorrect solution results from omitting $g_c$ in the equation.

(D) This answer results from erroneously assuming that vapor compression is an adiabatic process.

# Depth Solutions

## PRINCIPLES

### SOLUTION 21

When a cylinder is placed in a flowing liquid, it will experience a phenomenon known as vortex shedding. In the flow pattern around the cylinder, vortices are formed and shed from alternating sides of the cylinder. This formation and shedding of vortices from alternating sides creates a regular change in the pressures and subsequent forces acting on the sides of the cylinder. It is this change in side forces that causes the mine to sway back and forth in a plane perpendicular to the current velocity.

The frequency of the vortex shedding is given in terms of the Strouhal number, Sl, which is in turn a function of the Reynolds number. The Strouhal number is defined as

$$\text{Sl} = \frac{fd}{v_0}$$

The Strouhal number is obtained from a graph of Strouhal versus Reynolds numbers for flow past a cylinder. The vortex shedding frequency, $f$, will be equal to the natural frequencies given in the statement of the problem. With the cylinder's diameter, $d$, given, the free-stream velocity $v_0$ is the quantity to be found.

To find the Strouhal number, the Reynolds number must be calculated. The Reynolds number is defined as

$$\text{Re} = \frac{v_0 d}{\nu}$$

To calculate the Reynolds number, a current velocity is needed. Looking at the solutions, a reasonable choice is 1.0 kt, which is 1.688 ft/sec. Using the kinematic viscosity of seawater at 50°F, the Reynolds number is

$$\text{Re} = \frac{\left(1.688 \, \frac{\text{ft}}{\text{sec}}\right)(23 \text{ in})\left(\frac{1 \text{ ft}}{12 \text{ in}}\right)}{1.5 \times 10^{-4} \, \frac{\text{ft}^2}{\text{sec}}} \approx 2.2 \times 10^4$$

The kinematic viscosity of regular water ($1.58 \times 10^{-5}$ ft²/sec) could have been used, as seawater and water have similar properties. From the graph of Strouhal versus Reynolds numbers for flow past a circular cylinder, the Strouhal number is equal to 0.2.

The velocities can now be calculated using the Strouhal number equation.

$$v_1 = \frac{f_1 d}{\text{Sl}}$$

$$= \left(\frac{\left(0.31 \, \frac{\text{rad}}{\text{sec}}\right)(23 \text{ in})}{0.2}\right)\left(\frac{1 \text{ cycle}}{2\pi \text{ rad}}\right)$$

$$\times \left(\frac{1 \text{ Hz}}{1 \, \frac{\text{cycle}}{\text{sec}}}\right)\left(\frac{1 \text{ ft}}{12 \text{ in}}\right)\left(\frac{1 \text{ kt}}{1.688 \, \frac{\text{ft}}{\text{sec}}}\right)$$

$$= 0.280 \text{ kt}$$

$$v_2 = \frac{f_2 d}{\text{Sl}}$$

$$= \left(\frac{\left(1.81 \, \frac{\text{rad}}{\text{sec}}\right)(23 \text{ in})}{0.2}\right)\left(\frac{1 \text{ cycle}}{2\pi \text{ rad}}\right)$$

$$\times \left(\frac{1 \text{ Hz}}{1 \, \frac{\text{cycle}}{\text{sec}}}\right)\left(\frac{1 \text{ ft}}{12 \text{ in}}\right)\left(\frac{1 \text{ kt}}{1.688 \, \frac{\text{ft}}{\text{sec}}}\right)$$

$$= 1.64 \text{ kt}$$

**The answer is (A).**

Why Other Options Are Wrong

(B) This incorrect solution results when feet per second are not converted to knots.

(C) This incorrect solution results from not converting $f_1$ and $f_2$ from radians per second to hertz.

(D) This incorrect solution results from not converting the diameter from inches to feet.

## SOLUTION 22

At least three solution methods are possible.

### Method 1

The pump's total dynamic head or pump head can be calculated using the Bernoulli equation.

$$\frac{p_1}{\gamma} + \frac{v_1^2}{2g} + z_1 + h_A = \frac{p_2}{\gamma} + \frac{v_2^2}{2g} + z_2 + h_f$$

With the reservoir and tank surfaces as points 1 and 2, respectively, the gauge pressures and velocities at the surfaces are zero.

$$p_1 = v_1 = p_2 = v_2 = 0$$

The Bernoulli equation now becomes

$$0 + 0 + z_1 + h_A = 0 + 0 + z_2 + h_f$$
$$35 \text{ ft} + h_A = 150 \text{ ft} + h_f$$

$h_A$ is the pump's total dynamic head (TDH).

The head loss due to friction, $h_f$, can be found using the method of equivalent lengths. Since the suction and discharge lines are of different diameters, the friction head loss for each section must be calculated separately and then added together to find the total friction head loss.

### Suction

The equivalent length for a 10 in, flanged, schedule-40, steel gate valve is 3.2 ft. The equivalent suction line length is

$$L_{eq} = 200 \text{ ft} + (2)(3.2 \text{ ft}) = 206.4 \text{ ft}$$

The head loss per 100 linear feet of pipe, $h_f$, can be found using pipe friction tables. Be sure to use the water, schedule-40 steel pipe table. Looking up $h_f$ for a 10 in pipe with a flow rate of 1000 gpm, it is found that

$$h_f = \frac{0.5 \text{ ft}}{100 \text{ ft}}$$

The suction line head loss is

$$h_{f,S} = \left(\frac{0.5 \text{ ft}}{100 \text{ ft}}\right)(206.4 \text{ ft}) = 1.03 \text{ ft}$$

### Discharge

The equivalent length for an 8 in, flanged, schedule-40, steel gate valve is 3.2 ft. The equivalent length for an 8 in, flanged, schedule-40, steel swing valve is 90 ft. The equivalent length for an 8 in, flanged, schedule-40, regular steel 45° elbow is 7.7 ft. The equivalent discharge line length is

$$L_{eq} = 1500 \text{ ft} + (2)(3.2 \text{ ft}) + 90 \text{ ft} + (2)(7.7 \text{ ft})$$
$$= 1612 \text{ ft}$$

Using the pipe friction tables, $h_f$ is

$$h_f = \frac{1.56 \text{ ft}}{100 \text{ ft}}$$

The discharge line head loss is

$$h_{f,D} = \left(\frac{1.56 \text{ ft}}{100 \text{ ft}}\right)(1612 \text{ ft}) = 25.1 \text{ ft}$$

The total head loss is

$$h_f = h_{f,S} + h_{f,D} = 1.03 \text{ ft} + 25.1 \text{ ft} = 26.1 \text{ ft}$$

The pump's TDH can now be found.

$$35 \text{ ft} + h_A = 150 \text{ ft} + 26.1 \text{ ft}$$
$$h_A = 141.1 \text{ ft} \quad (140 \text{ ft})$$

*The answer is (A).*

### Method 2

The method of equivalent lengths is still used, but in this method the Darcy equation and the Moody friction factor, $f$, are used to determine the friction head losses. The Darcy equation is

$$h_f = \frac{f L_{eq} v^2}{2dg}$$

The suction and discharge lines are of different diameters, so the friction head loss for each section must be calculated separately and then added together to find the total friction head loss.

### Suction

The velocity can be calculated from the flow rate using the formula

$$Q = vA$$

$$\left(1000 \; \frac{\text{gal}}{\text{min}}\right)\left(\frac{1 \text{ min}}{60 \text{ sec}}\right) \times \left(\frac{1 \text{ ft}^3}{7.48 \text{ gal}}\right) = v\left(\frac{\pi(10.02 \text{ in})^2}{4}\right)\left(\frac{1 \text{ ft}}{12 \text{ in}}\right)^2$$

$$v = 4.07 \text{ ft/sec}$$

The Reynolds number is needed to find the Moody friction factor.

$$\text{Re} = \frac{vd}{\nu} = \frac{\left(4.07 \, \dfrac{\text{ft}}{\text{sec}}\right)(10.02 \text{ in})\left(\dfrac{1 \text{ ft}}{12 \text{ in}}\right)}{1.4 \times 10^{-5} \, \dfrac{\text{ft}^2}{\text{sec}}}$$

$$= 2.4 \times 10^5$$

The relative roughness is also needed (the specific roughness, $\varepsilon$, for steel is 0.0002 ft).

$$\frac{\varepsilon}{d} = \frac{0.0002 \text{ ft}}{(10.02 \text{ in})\left(\dfrac{1 \text{ ft}}{12 \text{ in}}\right)} = 0.00024$$

Using the Moody friction factor chart, the friction factor is

$$f = 0.017$$

The friction head loss in the suction line is

$$h_{f,S} = \frac{(0.017)(206.4 \text{ ft})\left(4.07 \, \dfrac{\text{ft}}{\text{sec}}\right)^2}{(2)(10.02 \text{ in})\left(\dfrac{1 \text{ ft}}{12 \text{ in}}\right)\left(32.2 \, \dfrac{\text{ft}}{\text{sec}^2}\right)} = 1.1 \text{ ft}$$

### Discharge

The velocity for the discharge line is found from the equation

$$Q = vA$$

$$\left(1000 \, \frac{\text{gal}}{\text{min}}\right)\left(\frac{1 \text{ min}}{60 \text{ sec}}\right)$$

$$\times \left(\frac{1 \text{ ft}^3}{7.48 \text{ gal}}\right) = v\left(\frac{\pi(7.981 \text{ in})^2}{4}\right)\left(\frac{1 \text{ ft}}{12 \text{ in}}\right)^2$$

$$v = 6.41 \text{ ft/sec}$$

The Reynolds number is

$$\text{Re} = \frac{vd}{\nu} = \frac{\left(6.41 \, \dfrac{\text{ft}}{\text{sec}}\right)(7.981 \text{ in})\left(\dfrac{1 \text{ ft}}{12 \text{ in}}\right)}{1.4 \times 10^{-5} \, \dfrac{\text{ft}^2}{\text{sec}}} = 3.0 \times 10^5$$

The relative roughness is

$$\frac{\varepsilon}{d} = \frac{0.0002 \text{ ft}}{(7.981 \text{ in})\left(\dfrac{1 \text{ ft}}{12 \text{ in}}\right)} = 0.0003$$

From the Moody friction factor chart, the friction factor is

$$f = 0.017$$

The friction head loss in the discharge line is

$$h_{f,D} = \frac{(0.017)(1612 \text{ ft})\left(6.41 \, \dfrac{\text{ft}}{\text{sec}}\right)^2}{(2)(7.981 \text{ in})\left(\dfrac{1 \text{ ft}}{12 \text{ in}}\right)\left(32.2 \, \dfrac{\text{ft}}{\text{sec}^2}\right)} = 26.3 \text{ ft}$$

The total head loss is

$$h_f = h_{f,S} + h_{f,D}$$
$$= 1.1 \text{ ft} + 26.3 \text{ ft}$$
$$= 27.4 \text{ ft}$$

The pump's TDH is

$$35 \text{ ft} + h_A = 150 \text{ ft} + 27.4 \text{ ft}$$
$$h_A = 142.4 \text{ ft} \quad (140 \text{ ft})$$

It should be noted that the Darcy friction factor table could also be used for $f$. Using a Darcy friction factor table, $f$ is 0.017 for both the suction and discharge lines.

**The answer is (A).**

### Method 3

If the pipe lengths are long and there are few fittings, the losses due to the fittings can be neglected to provide an approximation that can be used to select one of the answers.

The suction line head loss is

$$h_{f,S} = \left(\frac{0.5 \text{ ft}}{100 \text{ ft}}\right)(200 \text{ ft}) = 1.0 \text{ ft}$$

The discharge-line head loss is

$$h_{f,D} = \left(\frac{1.56 \text{ ft}}{100 \text{ ft}}\right)(1500 \text{ ft}) = 23.4 \text{ ft}$$

The pump's TDH is

$$35 \text{ ft} + h_A = 150 \text{ ft} + (1.0 \text{ ft} + 23.4 \text{ ft})$$
$$h_A = 139.4 \text{ ft} \quad (140 \text{ ft})$$

**The answer is (A).**

### Why Other Options Are Wrong

(B) This answer is incorrect because the height of the open tank is not equal to the TDH.

(C) This answer is incorrect because the 35 ft from the left side of the Bernoulli equation was not subtracted from the right side of the equation.

(D) This answer is incorrect because the 35 ft from the left side of the Bernoulli equation was added to, not subtracted from, the right side of the equation.

## SOLUTION 23

The needle valve in the line to the manifold controls the pressure to the manifold. The other needle valve allows oil to leave the line to the manifold and return to the pump intake line; therefore, this valve primarily controls the amount of oil flowing to the manifold.

**The answer is (B).**

### Why Other Options Are Wrong

(A) This answer is incorrect because neither needle valve primarily controls the pressure at the pump intake.

(C) This answer is incorrect because neither needle valve primarily controls the flow rate to the pump intake.

(D) This answer is incorrect because neither needle valve primarily controls the flow rate to the pump intake.

## SOLUTION 24

The force needed to overcome the moment of the spring steel can be found from the following equation ($d$ is the moment arm).

$$M = Fd$$

$$F = \frac{M}{d}$$

$$= \frac{0.5 \text{ in-lbf}}{2 \text{ in}}$$

$$= 0.25 \text{ lbf}$$

The force, $F$, comes from the weight of the water. The height, $h$, of the water needed to generate 0.25 lbf is found from the equation

$$W = \gamma V$$

$$0.25 \text{ lbf} = \left(62.4 \ \frac{\text{lbf}}{\text{ft}^3}\right)(3 \text{ in})(4 \text{ in})\left(\frac{1 \text{ ft}}{12 \text{ in}}\right)^3 h$$

$$h = 0.5769 \text{ in} \quad (0.58 \text{ in})$$

**The answer is (B).**

### Why Other Options Are Wrong

(A) This incorrect answer results from using the diameter instead of the radius when calculating the force.

(C) This incorrect answer results from using the moment instead of the force to calculate $h$.

(D) This incorrect answer results from placing an extra factor of 12 in the unit conversion.

## SOLUTION 25

The heat transfer rate from one fin is

$$Q = \eta_f h A_f (T_b - T_\infty)$$

To find the fin area, $A_f$, the corrected fin length is needed.

$$L_{\text{corr}} = L + \frac{t}{2}$$

$$= 1.25 \text{ in} + \frac{0.125 \text{ in}}{2}$$

$$= 1.3125 \text{ in}$$

Now the fin area can be calculated.

$$A_f = 2\pi\left((L_{\text{corr}} + r_1)^2 - r_1^2\right)$$

$$= 2\pi\left((1.3125 \text{ in} + 0.25 \text{ in})^2 - (0.25 \text{ in})^2\right)\left(\frac{1 \text{ ft}}{12 \text{ in}}\right)^2$$

$$= 0.104 \text{ ft}^2$$

The fin efficiency is found using a fin efficiency graph. The graph parameters are

$$\frac{L_{\text{corr}} + r_1}{r_1} = \frac{1.3125 \text{ in} + 0.25 \text{ in}}{0.25 \text{ in}} = 6.25$$

$$L_{\text{corr}} \sqrt{\frac{h}{kt}} = (1.3125 \text{ in})\left(\frac{1 \text{ ft}}{12 \text{ in}}\right)$$

$$\times \sqrt{\frac{3 \ \frac{\text{Btu}}{\text{hr-ft}^2\text{-}°\text{F}}}{\left(27 \ \frac{\text{Btu}}{\text{hr-ft-}°\text{F}}\right)(0.125 \text{ in})\left(\frac{1 \text{ ft}}{12 \text{ in}}\right)}}$$

$$= 0.357$$

From the graph, the fin efficiency is approximately

$$\eta_f = 0.8$$

The heat transfer rate per fin can now be found.

$$Q_{\text{fin}} = \eta_f h A_f (T_b - T_\infty)$$

$$= (0.8)\left(3 \ \frac{\text{Btu}}{\text{hr-ft}^2\text{-}°\text{F}}\right)(0.104 \text{ ft}^2)(150°\text{F} - 80°\text{F})$$

$$= 17.47 \text{ Btu/hr}$$

The total heat transfer rate of all 100 fins is

$$Q_{\text{total}} = (\text{no. of fins})Q_{\text{fin}}$$

$$= (100 \text{ fins})\left(17.47 \frac{\text{Btu}}{\text{hr-fin}}\right)$$

$$= 1747 \text{ Btu/hr} \quad (1700 \text{ Btu/hr})$$

**The answer is (C).**

### Why Other Options Are Wrong

(A) This incorrect answer results from neglecting the inches-to-feet conversion in calculating one of the graph parameters.

(B) This incorrect answer results from using $2\pi L_{\text{corr}}^2$ for the fin area.

(D) This incorrect answer results from neglecting to use the fin efficiency.

## SOLUTION 26

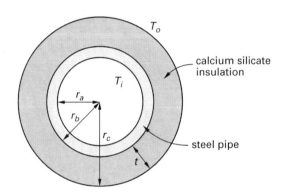

The heat transfer through the pipe is given by the equation

$$Q = \frac{2\pi L(T_i - T_o)}{\dfrac{1}{r_a h_a} + \dfrac{\ln\dfrac{r_b}{r_a}}{k_{\text{pipe}}} + \dfrac{1}{r_b h_b} + \dfrac{\ln\dfrac{r_c}{r_b}}{k_{\text{insul}}} + \dfrac{1}{r_c h_c}}$$

The outer radius of a 10 in schedule-80 pipe is

$$r_b = \frac{d_o}{2} = \frac{10.75 \text{ in}}{2} = 5.375 \text{ in}$$

The temperature drop in the pipe wall is negligible, and all film coefficients are infinite, so the heat loss through the pipe is

$$Q = \frac{2\pi L(T_i - T_o)k_{\text{insul}}}{\ln\dfrac{r_c}{r_b}}$$

The outer radius of the insulation is

$$r_c = r_b + t$$

Therefore,

$$\frac{r_c}{r_b} = \frac{r_b + t}{r_b}$$

$$= 1 + \frac{t}{r_b}$$

Substituting and solving for the thickness gives

$$Q = \frac{2\pi L(T_i - T_o)k_{\text{insul}}}{\ln\left(1 + \dfrac{t}{r_b}\right)}$$

$$t = r_b \left(e^{\frac{2\pi L(T_i - T_o)}{Q}k_{\text{insul}}} - 1\right)$$

The heat rate required to reduce the water temperature by 1°F is

$$Q = \dot{m}c_p \Delta T$$

$$= \left(7230 \frac{\text{lbm}}{\text{hr}}\right)\left(1.08 \frac{\text{Btu}}{\text{lbm-°F}}\right)(1°\text{F})$$

$$= 7810 \text{ Btu/hr}$$

The required thickness is

$$t = (5.375 \text{ in})\left(e^{\dfrac{(2\pi)(40 \text{ ft})(400°\text{F}-66°\text{F})\left(0.046 \frac{\text{Btu-ft}}{\text{hr-ft}^2\text{-°F}}\right)}{7810 \frac{\text{Btu}}{\text{hr}}}} - 1\right)$$

$$= 3.44 \text{ in} \quad (3.4 \text{ in})$$

**The answer is (A).**

### Why Other Options Are Wrong

(B) This incorrect answer results when a specific heat equal to 1.00 is used instead of 1.08.

(C) This incorrect answer results when the diameter is used instead of the radius in calculating the thickness.

(D) This incorrect answer results when the subtraction of the 1 in thickness is omitted.

## SOLUTION 27

The mass flow rate of vapor is

$$\dot{m} = A\alpha\left(\frac{\rho_0 - \rho_L}{L}\right)$$

$\rho_0$ is vapor density at $x = 0$ ft, and $\rho_L$ is vapor density at $x = L$. At each location, the vapor density is

$$\rho = \phi\rho_{\text{sat}}$$

From steam tables, the saturated vapor densities at the given temperatures are

$$\rho_{sat,70°} = 1.15 \times 10^{-3} \text{ lbm/ft}^3$$
$$\rho_{sat,60°} = 0.829 \times 10^{-3} \text{ lbm/ft}^3$$

Therefore,

$$\rho_0 = (0.80)\left(1.15 \times 10^{-3} \frac{\text{lbm}}{\text{ft}^3}\right)$$
$$= 0.922 \times 10^{-3} \text{ lbm/ft}^3$$
$$\rho_L = (0.10)\left(0.829 \times 10^{-3} \frac{\text{lbm}}{\text{ft}^3}\right)$$
$$= 0.829 \times 10^{-4} \text{ lbm/ft}^3$$

Substituting into the equation for mass flow rate,

$$\dot{m} = (2 \text{ ft}^2)\left(1.0 \frac{\text{ft}^2}{\text{hr}}\right)$$
$$\times \left(\frac{0.922 \times 10^{-3} \frac{\text{lbm}}{\text{ft}^3} - 0.829 \times 10^{-4} \frac{\text{lbm}}{\text{ft}^3}}{50 \text{ ft}}\right)$$
$$= 3.36 \times 10^{-5} \text{ lbm/hr} \quad (3.4 \times 10^{-5} \text{ lbm/hr})$$

**The answer is (C).**

Why Other Options Are Wrong

(A) This incorrect answer results when relative humidity is neglected.

(B) This incorrect answer results when the cross-sectional area of the duct is omitted from the calculation.

(D) This incorrect answer results when the relative humidity at $x = 0$ is neglected.

## SOLUTION 28

The vapor mass flow rate is

$$\dot{m} = h_m A(\rho_s - \rho_\infty)$$

The vapor density at the water surface is

$$\rho_s = \rho_{sat,60°} = 0.829 \times 10^{-3} \text{ lbm/ft}^3$$

The vapor density in the free stream is

$$\rho_\infty = \phi \rho_{sat,70°} = (0.50)\left(1.15 \times 10^{-3} \frac{\text{lbm}}{\text{ft}^3}\right)$$
$$= 0.575 \times 10^{-3} \text{ lbm/ft}^3$$

Thus,

$$\dot{m} = \left(0.10 \frac{\text{ft}}{\text{sec}}\right)(4 \text{ ft}^2)\left(\begin{array}{c} 0.829 \times 10^{-3} \frac{\text{lbm}}{\text{ft}^3} \\ - 0.575 \times 10^{-3} \frac{\text{lbm}}{\text{ft}^3} \end{array}\right)$$
$$\times \left(3600 \frac{\text{sec}}{\text{hr}}\right)$$
$$= 0.365 \text{ lbm/hr} \quad (0.37 \text{ lbm/hr})$$

**The answer is (B).**

Why Other Options Are Wrong

(A) This incorrect answer results when the area is not included in the calculation.

(B) This incorrect answer results when the relative humidity is neglected and the vapor densities are reversed.

(D) This incorrect answer results when the temperatures of air and water are switched.

## SOLUTION 29

First determine the brake horsepower needed for the original pump design. Using the performance curves, the brake horsepower for 350 gpm and 200 ft of head is approximately 28 bhp.

Now determine the brake horsepower for the oversized pump. The oversized pump rating is

$$Q_{os} = (\text{oversize factor})Q = (1.2)\left(350 \frac{\text{gal}}{\text{min}}\right)$$
$$= 420 \text{ gpm}$$
$$h_{os} = (\text{oversize factor})h = (1.2)(200 \text{ ft})$$
$$= 240 \text{ ft}$$

Using the performance curves, the brake horsepower for 420 gpm and 240 ft of head is approximately 40 bhp.

The larger pump will have to be throttled back to a flow rate of 350 gpm once put into service. Both pumps deliver 350 gpm, but the oversized pump needs an additional 12 hp to do it. This 12 hp is wasted energy, and the additional yearly electrical cost is

$$(12 \text{ hp})\left(0.746 \frac{\text{kW}}{\text{hp}}\right)\left(\frac{\$0.07}{\text{kW-hr}}\right)\left(24 \frac{\text{hr}}{\text{day}}\right)\left(365 \frac{\text{days}}{\text{yr}}\right)$$
$$= \$5489/\text{yr} \quad (\$5490/\text{yr})$$

**The answer is (D).**

## Why Other Options Are Wrong

(A) For this pump, a larger impeller diameter is needed to handle the 20% additional load. Thus, more power is needed, so the pump as originally designed cannot handle the additional load.

(B) This incorrect answer results when the brake horsepower of the original design is misread as 38 bhp.

(C) This incorrect answer results when, in oversizing the pump, only the flow rate is increased and the head increase is forgotten.

## SOLUTION 30

The straight-line depreciation for each year is given by

$$D = \frac{C - S_n}{n} = \frac{\$130{,}000 - \$10{,}000}{4} = \$30{,}000$$

Taxes are owed on net profits less depreciation, so

$$T = r_{\text{tax}}(P - D) = (0.45)(\$45{,}000 - \$30{,}000) = \$6750$$

Then the net income for each year is

$$\text{NI} = P - T = \$45{,}000 - \$6750 = \$38{,}250$$

The present worth of the equipment is the sum of the present worths of the net annual income and the salvage value, less the cost.

$$\begin{aligned}
P_{\text{income}} &= (\text{NI})(P/A, i, n) \\
&= (\$38{,}250)(P/A, 8\%, 4) \\
&= (\$38{,}250)(3.3121) \\
&= \$126{,}688 \\
P_{\text{salvage}} &= S(P/F, i, n) \\
&= (\$10{,}000)(P/F, 8\%, 4) \\
&= (\$10{,}000)(0.7350) \\
&= \$7350 \\
P &= P_{\text{income}} + P_{\text{salvage}} - C \\
&= \$126{,}688 + \$7350 - \$130{,}000 \\
&= \$4038 \ (\$4000)
\end{aligned}$$

The present value is greater than zero, so the purchase should be made.

**The answer is (C).**

### Why Other Options Are Wrong

(A) This incorrect answer results when the future value is calculated instead of the present value, and the signs are reversed.

(B) This incorrect answer results when the signs are reversed.

(D) This incorrect answer results when the future value is calculated instead of the present value.

## SOLUTION 31

### Method 1

The change in pressure can be found using the equation

$$\Delta p = \rho h_f \left(\frac{g}{g_c}\right) = \left(\frac{h_f}{v}\right)\left(\frac{g}{g_c}\right)$$

The head loss due to friction is

$$h_f = \frac{fLv^2}{2dg}$$

The friction factor, $f$, is found using a friction factor chart. To use this chart, however, the Reynolds number is needed, and to compute the Reynolds number it is necessary to have the viscosity of the steam. Since the viscosity of steam is difficult to find, a typical friction factor for turbulent flow in steel pipe of 0.02 will be used.

The steam velocity is

$$\text{v} = \frac{Q}{A} = \frac{\dot{m}}{\rho A} = \frac{\dot{m} v}{A}$$

From the steam tables the specific volume is

$$v = 1.0792 \text{ ft}^3/\text{lbm}$$

6 in schedule-40 pipe has an internal diameter and an internal area of

$$d_i = 6.065 \text{ in}$$

$$A = \pi \left(\frac{d_i}{2}\right)^2 = \pi \left(\frac{6.065 \text{ in}}{2}\right)^2 = 28.89 \text{ in}^2$$

Therefore,

$$\text{v} = \frac{\dot{m}v}{A} = \frac{\left(50{,}000 \ \frac{\text{lbm}}{\text{hr}}\right)\left(\frac{1 \text{ hr}}{3600 \text{ sec}}\right)\left(1.0792 \ \frac{\text{ft}^3}{\text{lbm}}\right)}{(28.89 \text{ in}^2)\left(\frac{1 \text{ ft}}{12 \text{ in}}\right)^2}$$

$$= 74.7 \text{ ft/sec}$$

The head loss due to friction is

$$h_f = \frac{fLv^2}{2dg} = \frac{(0.02)(100 \text{ ft})\left(74.7 \ \frac{\text{ft}}{\text{sec}}\right)^2}{(2)(6.065 \text{ in})\left(\frac{1 \text{ ft}}{12 \text{ in}}\right)\left(32.2 \ \frac{\text{ft}}{\text{sec}^2}\right)}$$

$$= 343 \text{ ft}$$

The pressure drop is

$$\Delta p = \left(\frac{h_f}{v}\right)\left(\frac{g}{g_c}\right)$$

$$= \left(\frac{(343\text{ ft})\left(12\,\frac{\text{in}}{\text{ft}}\right)}{\left(1.0792\,\frac{\text{ft}^3}{\text{lbm}}\right)\left(12\,\frac{\text{in}}{\text{ft}}\right)^3}\right)\left(\frac{32.2\,\frac{\text{ft}}{\text{sec}^2}}{32.2\,\frac{\text{ft-lbm}}{\text{lbf-sec}^2}}\right)$$

$$= 2.2\text{ lbf/in}^2$$

**The answer is (A).**

## Method 2

Steam viscosity data can be found in the Crane Company's technical paper no. 410. For steam at 550°F and 500 psia, the viscosity in centipoises is

$$\mu = 0.02\text{ cP}$$

Converting the viscosity to English units gives

$$\mu = (0.02\text{ cP})\left(\frac{1\text{ P}}{100\text{ cP}}\right)\left(0.002089\,\frac{\frac{\text{lbf-sec}}{\text{ft}^2}}{\text{P}}\right)$$

$$= 4.18 \times 10^{-7}\text{ lbf-sec/ft}^2$$

Now the Moody friction factor chart can be used to find the friction factor. To use this chart, the Reynolds number and the relative roughness are needed. When the mass flow rate is given, the Reynolds number can be expressed in terms of the mass flow rate per unit area, $G$.

$$G = \rho v = \frac{\dot{m}}{A} = \frac{\left(50{,}000\,\frac{\text{lbm}}{\text{hr}}\right)\left(\frac{1\text{ hr}}{3600\text{ sec}}\right)}{(28.89\text{ in}^2)\left(\frac{1\text{ ft}}{12\text{ in}}\right)^2}$$

$$= 69.2\text{ lbm/ft}^2\text{-sec}$$

$$\text{Re} = \frac{dG}{g_c\mu}$$

$$= \frac{(6.065\text{ in})\left(\frac{1\text{ ft}}{12\text{ in}}\right)\left(69.2\,\frac{\text{lbm}}{\text{ft}^2\text{-sec}}\right)}{\left(32.2\,\frac{\text{ft-lbm}}{\text{lbf-sec}^2}\right)\left(4.18\times 10^{-7}\,\frac{\text{lbf-sec}}{\text{ft}^2}\right)}$$

$$= 2.6 \times 10^6$$

The relative roughness is given by $\varepsilon/d$. The specific roughness, $\varepsilon$, can be found from tables, and for steel, $\varepsilon$ equals 0.0002 ft. The relative roughness is

$$\frac{\varepsilon}{d} = \frac{0.0002\text{ ft}}{(6.065\text{ in})\left(\frac{1\text{ ft}}{12\text{ in}}\right)} = 4\times 10^{-4}$$

From the Moody friction factor chart, the friction factor is

$$f = 0.016$$

The head loss due to friction is

$$h_f = \frac{fLv^2}{2dg} = \frac{(0.016)(100\text{ ft})\left(74.7\,\frac{\text{ft}}{\text{sec}}\right)^2}{(2)(6.065\text{ in})\left(\frac{1\text{ ft}}{12\text{ in}}\right)\left(32.2\,\frac{\text{ft}}{\text{sec}^2}\right)}$$

$$= 274\text{ ft}$$

The pressure drop is

$$\Delta p = \left(\frac{h_f}{v}\right)\left(\frac{g}{g_c}\right)$$

$$= \left(\frac{(274\text{ ft})\left(12\,\frac{\text{in}}{\text{ft}}\right)}{\left(1.0792\,\frac{\text{ft}^3}{\text{lbm}}\right)\left(12\,\frac{\text{in}}{\text{ft}}\right)^3}\right)\left(\frac{32.2\,\frac{\text{ft}}{\text{sec}^2}}{32.2\,\frac{\text{ft-lbm}}{\text{lbf-sec}^2}}\right)$$

$$= 1.76\text{ lbf/in}^2 \quad (2.2\text{ lbf/in}^2)$$

**The answer is (A).**

## Method 3

The Crane Company's technical paper no. 410 also gives the following equation for calculating the pressure drop. This combines all the previous equations and conversion factors into one equation.

$$\Delta p = 0.000336\left(\frac{f\dot{m}^2 v}{d^5}\right)$$

For this equation to be dimensionally correct, the constant must have the following units.

$$\Delta p = \left(0.000336\,\frac{\text{lbf-in}^3\text{-hr}^2}{\text{lbm-ft}^3}\right)\left(\frac{f\dot{m}^2 v}{d^5}\right)$$

To use this equation, the mass flow rate, $\dot{m}$, must be in lbm/hr, the specific volume, $v$, must be in ft$^3$/lbm, and the diameter, $d$, must be in inches.

$$\Delta p = \left(0.000336\,\frac{\text{lbf-in}^3\text{-hr}^2}{\text{lbm-ft}^3}\right)$$

$$\times\left(\frac{(0.02)\left(50{,}000\,\frac{\text{lbm}}{\text{hr}}\right)^2\left(1.0792\,\frac{\text{ft}^3}{\text{lbm}}\right)}{(6.065\text{ in})^5}\right)$$

$$= 2.2\text{ lbf/in}^2$$

(Using $f = 0.018$ in this equation yields $\Delta p = 2.0$ psi.)

**The answer is (A).**

## Method 4

The Babcock formula for pressure drop when steam flows in a pipe is

$$\Delta p_{psi} = (0.470)\left(\frac{d_{in} + 3.6}{d_{in}^6}\right)(\dot{m}_{lbm/sec})^2 L_{ft} v_{ft^3/lbm}$$

$$= (0.470)\left(\frac{6.065 + 3.6}{(6.065)^6}\right)\left(\frac{50{,}000}{3600}\right)^2$$
$$\quad \times (100)(1.0792)$$
$$= 1.9$$
$$\Delta p = 1.9 \text{ psi} \quad (2.2 \text{ lbf/in}^2)$$

## Method 5

The Crane Company's technical paper no. 410 also has a nomograph that can be used to solve this problem graphically. To use their nomograph follow these steps.

1. Locate 50 on the $W$ axis, noting that the scale is in 1000 lbm/hr. Label this point A.
2. Locate the nominal diameter of 6 in on the $d$ axis, and label this point B.
3. Draw a straight line through points A and B. Find the intersection of this line and the Index 2 axis, and label it point C.
4. Locate 0.02 on the friction factor $f$ axis and label it point D.
5. Draw a straight line between points C and D. Find the intersection of this line and the Index 1 axis, and label it point E.
6. Locate 1.0792 ft$^3$/lbm on the specific volume axis, and label this point F.
7. Draw a straight line between points E and F. The intersection of this line and the $\Delta P_{100}$ axis is the change in pressure, 2 psi.

**The answer is (A).**

### Why Other Options Are Wrong

(B) This answer results when the wrong internal pipe diameter is read from the schedule-40 table.

(C) This incorrect answer results when a specific volume of 1.792 ft$^3$/lbm, instead of 1.0792 ft$^3$/lbm, is used in the velocity calculation.

(D) This incorrect answer results when the conversion factor for $v$ is squared instead of cubed.

## SOLUTION 32

The flow is isentropic up to and after the normal shock. From the normal shock table, for $M_x$ equal to 1.8, $M_y$ is equal to 0.62. To determine the properties of the isentropic flow from there to the exit, the ratio of the exit area to the area at which the flow would attain Mach 1 must be known. Note that before the shock, this area $(A_x^*)$ is the throat area. After the shock, this area $(A_y^*)$ is larger. From the shock tables again, for $M_x$ equal to 1.8,

$$\frac{A_x^*}{A_y^*} = \frac{p_{0,y}}{p_{0,x}} = 0.81$$

The needed ratio, $A_{exit}/A_y^*$, can be calculated as follows.

$$\frac{A_{exit}}{A_y^*} = \left(\frac{A_{exit}}{A_x^*}\right)\left(\frac{A_x^*}{A_y^*}\right) = \left(\frac{A_{exit}}{A_{throat}}\right)\left(\frac{A_x^*}{A_y^*}\right)$$
$$= (3)(0.81)$$
$$= 2.4$$

From the isentropic tables, this ratio corresponds to a Mach number of 0.25.

**The answer is (B).**

### Why Other Options Are Wrong

(A) This answer is incorrect because the change in $A^*$ across the shock is not accounted for.

(C) This incorrect answer is the supersonic solution to the area ratio.

(D) This incorrect answer results when the change in $A^*$ across the shock is not accounted for and the supersonic solution is found.

## SOLUTION 33

The acceleration in the $x$-direction can be found with the equation

$$a_x = \frac{F_x g_c}{m_{block}}$$

Since $m_{block}$ is known, $F_x$ is needed. There are two forces in the $x$-direction: one due to the water, and the other due to friction.

The force that the block and vane exert on the water is given by the formula

$$F_{x,block} = \frac{\dot{m}_{water}(v_2 \cos\theta - v_1)}{g_c}$$

The mass flow rate is given by

$$\dot{m}_{\text{water}} = \rho v A$$
$$= \left(62.4 \ \frac{\text{lbm}}{\text{ft}^3}\right)\left(150 \ \frac{\text{ft}}{\text{sec}}\right)(0.005 \ \text{ft}^2)$$
$$= 46.8 \ \text{lbm/sec}$$

The velocity into and out of the vane is the same.

$$v_2 = v_1 = 150 \ \text{ft/sec}$$

Substituting,

$$F_{x,\text{block}} = \frac{\left(46.8 \ \dfrac{\text{lbm}}{\text{sec}}\right) \times \left(\left(150 \ \dfrac{\text{ft}}{\text{sec}}\right)\cos 30° - 150 \ \dfrac{\text{ft}}{\text{sec}}\right)}{32.2 \ \dfrac{\text{ft-lbm}}{\text{lbf-sec}^2}}$$
$$= -29.2 \ \text{lbf}$$

The force that the water exerts on the block and vane is equal and opposite to this.

$$F_{x,\text{water}} = -F_{x,\text{block}}$$
$$= 29.2 \ \text{lbf}$$

For the frictional force, the $y$-force due to the water is needed.

$$F_{y,\text{water}} = \frac{\dot{m}_{\text{water}} v_2 \sin\theta}{g_c}$$
$$= \frac{\left(46.8 \ \dfrac{\text{lbm}}{\text{sec}}\right)\left(150 \ \dfrac{\text{ft}}{\text{sec}}\right)\sin 30°}{32.2 \ \dfrac{\text{ft-lbm}}{\text{lbf-sec}^2}}$$
$$= 109 \ \text{lbf}$$

There is also the $y$-force due to the weight of the vane and block.

$$F_{y,\text{block}} = \frac{m_{\text{block}} g}{g_c}$$
$$= \frac{(50 \ \text{lbm})\left(32.2 \ \dfrac{\text{ft}}{\text{sec}^2}\right)}{32.2 \ \dfrac{\text{ft-lbm}}{\text{lbf-sec}^2}}$$
$$= 50 \ \text{lbf}$$

The total force in the vertical direction is

$$F_y = F_{y,\text{water}} + F_{y,\text{block}} = 109 \ \text{lbf} + 50 \ \text{lbf} = 159 \ \text{lbf}$$

The $x$-force due to friction can be found from the equation

$$F_{x,\text{friction}} = \mu_f F_y = (0.1)(159 \ \text{lbf}) = 15.9 \ \text{lbf}$$

Friction acts against the water force, so the resulting acceleration is

$$a_x = \frac{F_x g_c}{m} = \frac{(F_{x,\text{water}} - F_{x,\text{friction}}) g_c}{m_{\text{block}}}$$
$$= \frac{(29.2 \ \text{lbf} - 15.9 \ \text{lbf})\left(32.2 \ \dfrac{\text{ft-lbm}}{\text{lbf-sec}^2}\right)}{50 \ \text{lbm}}$$
$$= 8.57 \ \text{ft/sec}^2 \quad (8.6 \ \text{ft/sec}^2)$$

**The answer is (B).**

Why Other Options Are Wrong

(A) This incorrect answer results when the gravitational constant, $g_c$, is neglected.

(C) This incorrect answer results when friction is neglected.

(D) This incorrect answer results when the frictional force is added to the water force instead of subtracted from it.

### SOLUTION 34

Plot both cycles on the same $T$-$s$ diagram.

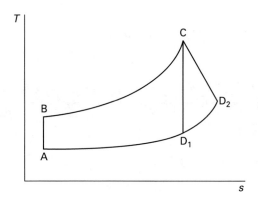

Thermal efficiency is defined as

$$\eta_{\text{th}} = \frac{W_{\text{turbine}} - W_{\text{compressor}}}{Q_{\text{in}}}$$

$Q_{\text{in}}$ depends on $T_C$ and $T_B$, and $W_{\text{compressor}}$ depends on $T_A$ and $T_B$, so both values are the same for both cycles. $W_{\text{turbine}}$, however, is larger for system 1 (related to $T_D - T_C$), so the thermal efficiency of system 1 is higher, and statement I is true.

As just described, the work of the two compressors is the same, so statement II is true.

The heat rejected during cooling varies in proportion to $T_D - T_A$, which is larger for system 2. Thus, statement III is false.

The work out of the turbine varies in proportion to $T_C - T_D$, which is larger for system 1. Thus, statement IV is true.

**The answer is (C).**

Why Other Options Are Wrong

(A) This choice is wrong because statement IV is also true.

(B) This choice is wrong because statements I and II are true, and III is false.

(D) This choice is wrong because statement III is false.

## SOLUTION 35

Using $x$ as the steam quality, the circulation ratio, or CR, is defined as

$$\text{CR} = \frac{1}{x}$$

A circulation ratio of five gives a steam quality of 0.2, or 20%, meaning that 1 lbm of the steam-water mixture contains 0.2 lbm of steam.

**The answer is (D).**

Why Other Options Are Wrong

(A) Using the CR definition reveals that there is four times more water in the mixture than steam.

(B) Using the CR definition reveals that there is four times more water in the mixture than steam.

(C) The quality factor measures the amount of dry steam, not the amount of water in a steam-water mixture.

## SOLUTION 36

To estimate the maximum temperature, determine the adiabatic flame temperature.

$$T_{\max} = T_i + \frac{\text{lower heat of combustion}}{m_{\text{products}} c_{p,\text{mean}}}$$

The initial temperature, $T_i$, is 537°R.

Water vapor has a molecular weight of 18 lbm/lbmol. The mass of 2 lbmol of water vapor is 36 lbm.

The mass must be normalized by the mass of the fuel, to correspond to the normalization of the lower heat of combustion.

$$m_{\text{products}} = \frac{m_{\text{wv}}}{m_{\text{H}_2}} = \frac{36 \text{ lbm}}{4 \text{ lbm}_{\text{H}_2}} = 9 \text{ lbm/lbm}_{\text{H}_2}$$

The lower heat of combustion is 51,623 Btu/lbm$_{\text{H}_2}$.

The maximum temperature can now be found.

$$T_{\max} = T_i + \frac{\text{lower heat of combustion}}{m_{\text{products}} c_{p,\text{mean}}}$$

$$= 537°\text{R} + \frac{51{,}623 \dfrac{\text{Btu}}{\text{lbm}_{\text{H}_2}}}{\left(9 \dfrac{\text{lbm}}{\text{lbm}_{\text{H}_2}}\right)\left(1.15 \dfrac{\text{Btu}}{\text{lbm-°R}}\right)}$$

$$= 5520°\text{R} \quad (5500°\text{R})$$

**The answer is (D).**

Why Other Options Are Wrong

(A) This incorrect answer results when the product mass is not normalized and the initial temperature is omitted.

(B) This incorrect answer results when the product mass is not normalized by the mass of the fuel.

(C) This incorrect answer results when the initial temperature is omitted.

## SOLUTION 37

The balanced stoichiometric reaction must be found first. For this stoichiometric reaction, only $H_2O$, $CO_2$, and $N_2$ are allowed as products. Thus, for 1 mol of propane,

$$C_3H_8 + a(O_2 + 3.773N_2) \rightarrow bH_2O + cCO_2 + dN_2$$

From conservation of mass and the three carbon atoms on the left side, there must be three $CO_2$ molecules on the right side, and therefore $c$ equals 3. There are eight H atoms on the left, so there must be four $H_2O$ molecules on the right, and $b$ equals 4. There are then 10 atoms of oxygen on the right side, so there must be five $O_2$ molecules on the left, and $a$ equals 5. Finally, there are $5 \times 3.773$ or 18.87 molecules of $N_2$ per mole of propane on the left, so there must be 18.87 molecules of $N_2$ on the right, and $d$ equals 18.87. The balanced, stoichiometric equation is then

$$C_3H_8 + 5(O_2 + 3.773N_2) \rightarrow 4H_2O + 3CO_2 + 18.87N_2$$

Next, analyze the Orsat data to determine the actual reaction ratios. The reaction with the extended product list is

$$eC_3H_8 + f(O_2 + 3.773N_2)$$
$$\rightarrow gH_2O + hCO_2 + iO_2 + jCO + kN_2$$

For every 100 mol of Orsat products, there are 7.7 mol of $CO_2$, 6.6 mol of $O_2$, and 2.2 mol of CO. The remaining 83.5 mol must be $N_2$. (The Orsat analysis is a dry analysis and does not provide information about the water content.) Substituting,

$$eC_3H_8 + f(O_2 + 3.773N_2)$$
$$\rightarrow gH_2O + 7.7CO_2 + 6.6O_2 + 2.2CO + 83.5N_2$$

Conservation of mass can be used to determine the remaining unknowns. For carbon,

$$3e = 7.7 + 2.2$$
$$e = 3.3$$

Then, for hydrogen,

$$(8)(3.3) = 2g$$
$$g = 13.2$$

And then, for oxygen,

$$2f = 13.2 + 15.4 + 13.2 + 2.2$$
$$f = 22$$

Check the nitrogen balance. On the left are 166 nitrogen atoms. On the right are 167. Taking rounding errors into account, particularly with the Orsat analysis numbers, this is a good balance.

The final reaction is then

$$3.3C_3H_8 + 22(O_2 + 3.773N_2)$$
$$\rightarrow 13.2H_2O + 7.7CO_2 + 6.6O_2 + 2.2CO + 83.5N_2$$

In the stoichiometric reaction, there are 16.5 mol of air for every 3.3 mol of propane. There are 3.3 mol of propane and 22 mol of air in the actual reaction, so the excess air is

$$\text{excess air} = \frac{\text{actual air} - \text{stoichiometric air}}{\text{stoichiometric air}}$$
$$= \frac{22 \text{ mol} - 16.5 \text{ mol}}{16.5 \text{ mol}}$$
$$= 0.33 \quad (33\%)$$

**The answer is (C).**

Why Other Options Are Wrong

(A) This incorrect answer results when the CO and $CO_2$ values are reversed in the Orsat analysis.

(B) This incorrect answer results when the excess air is calculated relative to the actual air instead of to the stoichiometric air.

(D) This incorrect answer results when the stoichiometric reaction is mistakenly calculated with CO instead of $CO_2$.

## SOLUTION 38

The shear stress in the bolts is found using the average shear stress equation.

$$\tau = \frac{V}{A}$$

In this case the shear force, $V$, is the vertical reaction acting on the bolts. As shown, there is no significant velocity in the vertical direction where the flow passes across the control surface, so there is no change in momentum in this direction.

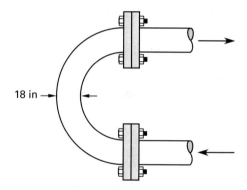

Consequently, the reaction that must be resisted by the bolts in the vertical direction is obtained by static equilibrium.

$$\sum F_{\text{vert}} = 0 \text{ lbf}$$
$$W_{\text{bend}} + W_{\text{fluid}} + R_{\text{bolts}} = 0 \text{ lbf}$$
$$627 \text{ lbf} + (100 \text{ gal})$$
$$\times \left(\frac{1 \text{ ft}^3}{7.48 \text{ gal}}\right)\left(62.4 \frac{\text{lbf}}{\text{ft}^3}\right)(0.94) + R_{\text{bolts}} = 0 \text{ lbf}$$
$$R_{\text{bolts}} = 1410 \text{ lbf}$$

Typically, the nominal diameter of the bolt is used to compute the shear area. The bend and the fluid are supported by 32 bolts (16 in each flange).

$$\tau = \frac{V}{A}$$
$$= \frac{1410 \text{ lbf}}{(0.25\pi)(1 \text{ in})^2(32 \text{ bolts})}$$
$$= 56 \text{ psi/bolt}$$

**The answer is (C).**

Note that bolt failure due to direct shear is uncommon. Bolt failure due to torsional and thread shear, however, is common, and engineers need to pay close attention to these modes of failure.

### Why Other Options Are Wrong

(A) This incorrect solution results from neglecting the weight of the fluid in the bend.

(B) This incorrect solution results from neglecting the weight of the bend.

(D) This incorrect solution results from using only 16 bolts to support bend and fluid weight.

## SOLUTION 39

This problem can be solved using the power and torsional stress equations.

$$P = T\omega$$
$$\tau = \frac{Tr}{J}$$

Using the maximum shear stress failure criterion and a yield stress of 30,000 psi for the steel shaft, the maximum allowable shear stress in the shaft is

$$\tau = 0.5\sigma_y = (0.5)\left(30{,}000 \ \frac{\text{lbf}}{\text{in}^2}\right) = 15{,}000 \text{ psi}$$

The maximum torque is found from the equation

$$\tau = \frac{Tr}{J}$$

$$15{,}000 \ \frac{\text{lbf}}{\text{in}^2} = \frac{T(1.25 \text{ in})}{\left(\frac{\pi}{32}\right)d^4} = \frac{T(1.25 \text{ in})}{\left(\frac{\pi}{32}\right)(2.5 \text{ in})^4}$$

$$T = 46{,}000 \text{ lbf-in}$$

The maximum power that can be transmitted is

$$P = T\omega$$
$$= (46{,}000 \text{ lbf-in})\left(\frac{1 \text{ ft}}{12 \text{ in}}\right)\left(200 \ \frac{\text{rev}}{\text{min}}\right)\left(2\pi \ \frac{\text{rad}}{\text{rev}}\right)$$
$$\times \left(\frac{1 \text{ min}}{60 \text{ sec}}\right)\left(\frac{1 \text{ hp}}{550 \ \frac{\text{lbf-ft}}{\text{sec}}}\right)$$
$$= 146 \text{ hp} \quad (150 \text{ hp})$$

**The answer is (B).**

### Why Other Options Are Wrong

(A) This incorrect answer results from not converting units of pound-feet per second to horsepower and adjusting the decimal point to bring the answer into a reasonable range.

(C) This incorrect answer results when the cross-sectional area is used instead of the polar area moment of inertia.

(D) This incorrect answer results when 30,000 psi is used as the maximum shear stress instead of 15,000 psi.

## SOLUTION 40

The band actually contacts the pipe slightly higher than its midheight. There is no short way to calculate the exact points of contact, but a good estimate can be made by solving the problem as though the band contacted the pipe at midheight exactly. The distance between these points is thus assumed to be the diameter of the pipe, which is given. This is slightly larger than the true distance between the points, so the result will be a slightly conservative (and therefore prudent) estimate of the maximum pressure.

The free-body diagram is as follows.

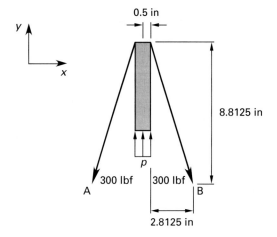

The $x$-component of the length of the tension in each side of the band is

$$L_x = \frac{d}{2} - \frac{w_{\text{nipple}}}{2}$$
$$= \frac{6.625 \text{ in}}{2} - \frac{1 \text{ in}}{2}$$
$$= 2.8125 \text{ in}$$

The $y$-component is

$$L_y = \frac{d}{2} + h_{\text{nipple}}$$
$$= \frac{6.625 \text{ in}}{2} + 5.5 \text{ in}$$
$$= 8.8125 \text{ in}$$

Summing forces in the $y$-direction,

$$\sum F_y = 0$$

$$F_p - (2)(300 \text{ lbf}) \times \left( \frac{8.8125 \text{ in}}{\sqrt{(8.8125 \text{ in})^2 + (2.8125 \text{ in})^2}} \right) = 0 \text{ lbf}$$

$$F_p = 572 \text{ lbf}$$

The maximum pressure is

$$p = \frac{F_p}{A} = \frac{572 \text{ lbf}}{\frac{\pi d^2}{4}} = \frac{572 \text{ lbf}}{\frac{(\pi)(1 \text{ in})^2}{4}}$$

$$= 728 \text{ psi} \quad (730 \text{ psi})$$

**The answer is (D).**

Why Other Options Are Wrong

(A) This incorrect solution results when, in the calculation of $F_p$, 2.8125 in is used in the numerator instead of 8.8125 in.

(B) This incorrect solution results when, in the calculation of $F_p$, the band force is not doubled.

(C) This incorrect solution results when the 100 lbf initial tension is subtracted from the 300 lbf breaking strength.

## SOLUTION 41

The bottom belt is gaining mass, so from momentum principles,

$$\Sigma F_x = m \left( \frac{dv_x}{dt} \right) + v_{\text{rel}} \left( \frac{dm_i}{dt} \right)$$

$v_{\text{rel}}$ is the relative velocity of the belt with respect to the particle. Since the belt is traveling at a constant velocity, the first term on the right side of the equation is equal to zero. The mass inflow rate is

$$\frac{dm_i}{dt} = \left( 77 \ \frac{\text{tons}}{\text{hr}} \right) \left( \frac{1 \text{ hr}}{3600 \text{ sec}} \right) \left( 2000 \ \frac{\text{lbf}}{\text{ton}} \right)$$

$$\times \left( \frac{1}{32.2 \ \frac{\text{ft}}{\text{sec}^2}} \right) \left( 1 \ \frac{\text{slug}}{\frac{\text{lbf-sec}^2}{\text{ft}}} \right)$$

$$= 1.33 \text{ slugs/sec}$$

Draw a free-body diagram of a portion of the belt that the mulch lands on, as shown.

Knowing that the horizontal velocity of the mulch just before it hits the lower belt is equal to the $x$ velocity at which the mulch leaves the upper belt, the initial equation is equivalent to

$$T_R - T_L = m \left( \frac{dv_x}{dt} \right) + v_{\text{rel}} \left( \frac{dm_i}{dt} \right)$$

$$= 0 + \left( 850 \ \frac{\text{ft}}{\text{min}} - 400 \ \frac{\text{ft}}{\text{min}} \right) \left( \frac{1 \text{ min}}{60 \text{ sec}} \right)$$

$$\times \left( 1.33 \ \frac{\text{slugs}}{\text{sec}} \right) \left( 1 \ \frac{\frac{\text{lbf-sec}^2}{\text{ft}}}{\text{slug}} \right)$$

$$= 9.98 \text{ lbf}$$

With the driver end of the lower conveyor system at the discharge side, the upper belt is at a higher tension than the lower belt, because the upper belt is being "pulled on" by the roller and the lower belt is being "pushed on" by the roller. The belt friction equation is

$$T_R = T_B e^{\mu \beta}$$

The tension in the bottom belt, $T_B$, is 100 lbf, so the tension can be found in the upper belt to the right of where the mulch lands.

$$T_R = (100 \text{ lbf}) e^{0.25\pi}$$

$$= 219 \text{ lbf}$$

The belt tension to the left of where the mulch lands is

$$T_R - T_L = 9.98 \text{ lbf}$$

$$T_L = T_R - 9.98 \text{ lbf}$$

$$= 219 \text{ lbf} - 9.98 \text{ lbf}$$

$$= 209 \text{ lbf} \quad (210 \text{ lbf})$$

**The answer is (D).**

Why Other Options Are Wrong

(A) This answer is incorrect because when calculating the upper belt tension from the belt friction equation, the 100 lbf tension was used in the left side of the equation instead of the right.

(B) This answer is incorrect because 13.3 slugs/sec was used instead of 1.33 slugs/sec.

(C) This answer is incorrect because the two velocities were added together to find a $v_{\text{rel}}$ of 1250 ft/min.

## SOLUTION 42

In conventional welding engineering practice, the external loading in a fillet weld is carried as a pure shear stress in the throat area. The shear stress in one fillet weld is given by

$$\tau = \frac{F}{0.707hL}$$

The throat area is $0.707hL$ for each weld. Since there are two welds, one on the inside and one on the outside, the average shear stress is

$$\tau = \frac{F}{1.414hL}$$

The allowable shear stress, $\tau$, is known, but the shear force, $F$, and length of the weld, $L$, are not. The length of the weld is the circumference of the center pipe.

$$L = \pi d = \pi(44 \text{ in}) = 138 \text{ in}$$

The force can be found by calculating the strength of the 38 bolts. Since the bolts are subjected to uniaxial stress, the maximum normal stress is

$$\sigma = \frac{F}{A}$$

The effective cross-sectional area for a ½ in UNF 20 bolt is 0.1599 in², and the tensile stress for an ASTM A354 grade BC bolt is 125 ksi.

The external load for each bolt is

$$\begin{aligned} F_{\text{bolt}} &= \sigma A \\ &= \left(125 \, \frac{\text{kip}}{\text{in}^2}\right)\left(1000 \, \frac{\text{lbf}}{\text{kip}}\right)(0.1599 \text{ in}^2) \\ &= 19{,}987.5 \text{ lbf} \end{aligned}$$

The total external load is

$$\begin{aligned} F_{\text{total}} &= F_{\text{bolt}}(\text{no. of bolts}) \\ &= (19{,}987.5 \text{ lbf})(38) \\ &= 759{,}525 \text{ lbf} \end{aligned}$$

The weld depth is found to be

$$\begin{aligned} h &= \frac{(\text{strength factor})F}{1.414L\tau} \\ &= \frac{(2)(759{,}525 \text{ lbf})}{(1.414)(138 \text{ in})\left(35 \, \frac{\text{kip}}{\text{in}^2}\right)\left(1000 \, \frac{\text{lbf}}{\text{kip}}\right)} \\ &= 0.222 \text{ in} \quad (0.250 \text{ in}) \end{aligned}$$

**The answer is (A).**

### Why Other Options Are Wrong

(B) This incorrect answer results when, in the calculation of $h$, the weld area is not doubled.

(C) This incorrect answer results when, in calculating the force on the bolts, the nominal cross-sectional area is used instead of the actual cross-sectional area, and the weld area is not doubled.

(D) This incorrect answer results when, in calculating the force on the bolts, the nominal cross-sectional area is used instead of the actual cross-sectional area, the weld area is not doubled, and an incorrect maximum bolt tensile stress of 150,000 psi is used.

## SOLUTION 43

Because the ends are fixed supports, this is a statically indeterminate problem.

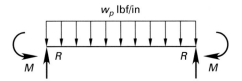

The load on the pipe consists of the weight of the pipe and the weight of the natural gas inside the pipe. Both loads can be modeled as uniform loads.

The weight per inch of pipe, $W_p$, is calculated as follows. The inner diameter of 30 in schedule-30 pipe is 28.75 in. The weight per inch of pipe is calculated by

$$\begin{aligned} W_p &= V\rho_p \\ &= AL\rho_p \\ &= \left(\frac{\pi(30 \text{ in})^2}{4} - \frac{\pi(28.75 \text{ in})^2}{4}\right)L\left(0.282 \, \frac{\text{lbf}}{\text{in}^3}\right) \\ &= \left(16.3 \, \frac{\text{lbf}}{\text{in}}\right)L \end{aligned}$$

$$\frac{W_p}{L} = w_{p,\text{inch}} = 16.3 \text{ lbf/in}$$

The weight per inch of natural gas inside the pipe is calculated by

$$\begin{aligned} W_g &= V\rho_g \\ &= AL\rho_g \\ &= \left(\frac{\pi(28.75 \text{ in})^2}{4}\right)L\left(0.046 \, \frac{\text{lbf}}{\text{ft}^3}\right)\left(\frac{1 \text{ ft}}{12 \text{ in}}\right)^3 \\ &= \left(0.017 \, \frac{\text{lbf}}{\text{in}}\right)L \end{aligned}$$

$$\frac{W_g}{L} = w_{g,\text{inch}} = 0.017 \text{ lbf/in}$$

The weight of the natural gas is small compared to the weight of the pipe and will be neglected. The equivalent beam solution is shown by the following calculations.

From the beam tables, the maximum moment is

$$M_{\max} = \frac{w_{p,\text{inch}} L^2}{12} = \frac{\left(16.3 \, \frac{\text{lbf}}{\text{in}}\right)\left((130 \text{ ft})\left(12 \, \frac{\text{in}}{\text{ft}}\right)\right)^2}{12}$$
$$= 3.3 \times 10^6 \text{ lbf-in}$$

The maximum bending stress is

$$\sigma_{\text{bend}} = \frac{M_{\max} c}{I} = \frac{(3.3 \times 10^6 \text{ lbf-in})(15 \text{ in})}{\left(\frac{\pi}{64}\right)((30 \text{ in})^4 - (28.75 \text{ in})^4)}$$
$$= 7.95 \times 10^3 \text{ lbf/in}^2 \quad (7.9 \text{ ksi})$$

**The answer is (B).**

Why Other Options Are Wrong

(A) This incorrect answer results when the polar area moment of inertia is used instead of the rectangular area moment of inertia.

(C) This incorrect answer results when the beam is modeled as a simply supported beam with a uniform load.

(D) This incorrect answer results when the beam is modeled as a simply supported beam with a concentrated load at the center.

## SOLUTION 44

The heat needed can be calculated using the formula

$$Q = \dot{m}\Delta h$$

From a psychrometric chart, the enthalpy, $h$, for 84°F air with 50% relative humidity is

$$h_{84} = 34.1 \text{ Btu/lbm}$$

The 102°F air must have the same humidity ratio as the 84°F air, because the moisture content is to remain unchanged. Again from the psychrometric chart, the humidity ratio is

$$\omega = 0.0125$$

From the psychrometric chart, the enthalpy for 102°F air with a 0.0125 humidity ratio is

$$h_{102} = 38.6 \text{ Btu/lbm}$$

The heat needed is

$$Q = \dot{m}\Delta h = \left(2500 \, \frac{\text{lbm}}{\text{hr}}\right)\left(38.6 \, \frac{\text{Btu}}{\text{lbm}} - 34.1 \, \frac{\text{Btu}}{\text{lbm}}\right)$$
$$= 11{,}250 \text{ Btu/hr} \quad (11 \times 10^3 \text{ Btu/hr})$$

**The answer is (B).**

Why Other Options Are Wrong

(A) This incorrect answer results when the 60% relative humidity line is used instead of the 50% relative humidity line.

(C) This incorrect answer results when a mass flow rate of 5200 lbm/hr, instead of 2500, is used.

(D) This incorrect answer results from incorrectly assuming that the relative humidity for both temperatures must be equal.

## SOLUTION 45

The root-locus sketch of the open-loop transfer function $G(s)$ shown here indicates that the closed-loop system is stable for all values of $K$.

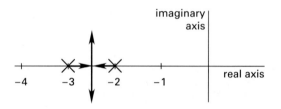

**The answer is (D).**

Why Other Options Are Wrong

(A) This incorrect answer is the value of the gain where the root locus breaks away from the negative real axis.

(B) This incorrect answer is the breakaway point on the negative real axis.

(C) This incorrect answer is part of the constant in the closed-loop characteristic equation.

## SOLUTION 46

The cycle thermal efficiency is

$$\eta_{\text{th}} = \frac{W_{\text{turbine}} - W_{\text{pump}}}{Q_{\text{in}}}$$

Neglecting pump work, this simplifies to

$$\eta_{\text{th}} = \frac{W_{\text{turbine}}}{Q_{\text{in}}}$$

The work of the turbine can be found from the turbine output and the mass flow rate.

$$W_{\text{turbine}} = \frac{W_{\text{out}}}{\dot{m}}$$

$$= \frac{(10\text{ MW})\left(10^6 \frac{\text{W}}{\text{MW}}\right)\left(3.413 \frac{\text{Btu}}{\text{hr-W}}\right)}{10^5 \frac{\text{lbm}}{\text{hr}}}$$

$$= 341.3 \text{ Btu/lbm}$$

The heat transferred into the system, again neglecting the work of the pump, equals the enthalpy of the saturated steam leaving the boiler minus the enthalpy of the saturated liquid leaving the condenser.

$$Q_{\text{in}} = h_{\text{steam}} - h_{\text{liquid}}$$

In the condenser, the condensate is subcooled 6°F. The saturation temperature for water at 2 psia is 126.04°F, so the temperature after subcooling is 120.04°F. From steam tables, the enthalpy of this saturated liquid is 88.00 Btu/lbm. Also from steam tables, the enthalpy of the saturated steam leaving the boiler at 1000 psia is 1192.4 Btu/lbm. Again neglecting pump work,

$$\eta_{\text{th}} = \frac{W_{\text{turbine}}}{Q_{\text{in}}} = \frac{W_{\text{turbine}}}{h_{\text{steam}} - h_{\text{liquid}}}$$

$$= \frac{341.3 \frac{\text{Btu}}{\text{lbm}}}{1192.4 \frac{\text{Btu}}{\text{lbm}} - 88.0 \frac{\text{Btu}}{\text{lbm}}}$$

$$= 0.31 \quad (31\%)$$

**The answer is (C).**

Why Other Options Are Wrong

(A) This incorrect solution results from leaving out the conversion from watts to Btu/hr.

(B) This incorrect solution results from adjusting the turbine work by multiplying by the turbine efficiency.

(D) This incorrect solution results from adjusting the turbine work by dividing by the turbine efficiency.

# EQUIPMENT

## SOLUTION 47

This problem is solved using the pump affinity laws.

$$\frac{Q_1}{Q_2} = \frac{n_1}{n_2}$$

$$\frac{h_1}{h_2} = \left(\frac{n_1}{n_2}\right)^2$$

With the given curves, the only way to solve this problem is through trial and error. The known values are

$$Q_2 = 550 \text{ gpm}$$
$$h_2 = 350 \text{ ft}$$
$$n_1 = 3550 \text{ rpm}$$

Since $n_2$ is not known, a speed must be guessed. If the resulting values for $Q_1$ and $h_1$ lie on the 9.5 in impeller curve, then the speed is correct. If not, a new guess must be tried.

Guess 1: $n_2$ is 3800 rpm.

$$Q_1 = \left(\frac{3550 \frac{\text{rev}}{\text{min}}}{3800 \frac{\text{rev}}{\text{min}}}\right)\left(550 \frac{\text{gal}}{\text{min}}\right) = 514 \text{ gpm}$$

$$h_1 = \left(\frac{3550 \frac{\text{rev}}{\text{min}}}{3800 \frac{\text{rev}}{\text{min}}}\right)^2 (350 \text{ ft}) = 305 \text{ ft}$$

This point lies near the $10^{3}/_{16}$ in diameter impeller curve, well above the 9.5 in diameter impeller curve.

Guess 2: $n_2$ is 4100 rpm.

$$Q_1 = \left(\frac{3550 \frac{\text{rev}}{\text{min}}}{4100 \frac{\text{rev}}{\text{min}}}\right)\left(550 \frac{\text{gal}}{\text{min}}\right) = 476 \text{ gpm}$$

$$h_1 = \left(\frac{3550 \frac{\text{rev}}{\text{min}}}{4100 \frac{\text{rev}}{\text{min}}}\right)^2 (350 \text{ ft}) = 262 \text{ ft}$$

This point lies close to the 9.5 in impeller curve, so the speed to produce 550 gpm and 350 ft of head is approximately 4100 rpm.

Guess 3: $n_2$ is 4150 rpm.

$$Q_1 = \left(\frac{3550 \frac{\text{rev}}{\text{min}}}{4150 \frac{\text{rev}}{\text{min}}}\right)\left(550 \frac{\text{gal}}{\text{min}}\right) = 470 \text{ gpm}$$

$$h_1 = \left(\frac{3550 \frac{\text{rev}}{\text{min}}}{4150 \frac{\text{rev}}{\text{min}}}\right)^2 (350 \text{ ft}) = 256 \text{ ft}$$

This point lies right on the 9.5 in impeller curve. So the required speed is 4150 rpm.

**The answer is (B).**

### Why Other Options Are Wrong

(A) This incorrect answer results when, in the calculation of $h_1$, the $n_1$ and $n_2$ speeds in the speed ratio are reversed.

(C) This incorrect answer results when, in the calculation of $h_1$, the speed ratio is not squared.

(D) This incorrect answer results when, in the calculation of $h_1$, 550 gpm is used instead of the 350 ft head.

### SOLUTION 48

The operating point of a pump is where the system curve crosses the pump curve. The system curve can be found, using the affinity laws, from the following equation and by developing a table of values for $Q_1$ versus $h_1$.

$$\frac{h_1}{h_2} = \left(\frac{Q_1}{Q_2}\right)^2$$

$$h_{1,\text{ft}} = h_{2,\text{ft}} \left(\frac{Q_{1,\text{gpm}}}{Q_{2,\text{gpm}}}\right)^2$$

$$= 200 \left(\frac{Q_{1,\text{gpm}}}{500}\right)^2$$

$$= (8 \times 10^{-4}) Q_{1,\text{gpm}}^2$$

| $Q_1$ (gpm) | $h_1$ (ft) |
|---|---|
| 0 | 0 |
| 200 | 32 |
| 400 | 128 |
| 600 | 288 |
| 800 | 512 |
| 1000 | 800 |

Plotting these points on the given graph, it can be seen that the system curve crosses the pump curve at approximately $Q = 800$ gpm. (A more exact answer is $Q = 785$ gpm.)

**The answer is (C).**

Why Other Options Are Wrong

(A) This incorrect answer results from switching 200 and 500 in the equation for $h_{1,\text{ft}}$.

(B) This incorrect answer results from using the point where the pump efficiency curve crosses the pump curve.

(D) This incorrect answer results from neglecting to square the ratio $Q_1/Q_2$.

### SOLUTION 49

The brake horsepower for the pump can be calculated using the two equations

$$\text{WHP} = \frac{h_A Q(\text{SG})}{3956}$$

$$\text{BHP} = \frac{\text{WHP}}{\eta} = \frac{h_A Q(\text{SG})}{3956\eta}$$

The total head is determined from the Bernoulli equation.

$$h_A = \frac{(p_D - p_S)g_c}{\rho g} + \frac{\text{v}_D^2 - \text{v}_S^2}{2g} + z_D - z_S + h_f$$

Since the static pressure head and the velocity heads are negligible, the expression for total head reduces to

$$h_A = z_D - z_S + h_f$$

The total friction head loss is given by

$$h_f = \frac{\Delta p_{\text{total}} g_c}{\rho g}$$

The total pressure drop due to friction in piping lines in series is the sum of the single pressure drops in the lines. The pressure drops in the two branches are combined using the equation

$$\Delta p_{\text{parallel}} = \frac{\Delta p_1 \Delta p_2}{\Delta p_1 + \Delta p_2}$$

The piping system consists of the series combination of the pump suction and discharge lines, and the parallel combination of the feed lines. Note that the pressure drop in the feed lines includes both pipe friction and minor valve losses. The total pressure drop in the piping system is therefore

$$\Delta p_{\text{total}} = \Delta p_S + \Delta p_D + \Delta p_{\text{parallel}}$$

$$= \Delta p_S + \Delta p_D + \frac{\Delta p_1 \Delta p_2}{\Delta p_1 + \Delta p_2}$$

$$= 23.2 \, \frac{\text{lbf}}{\text{in}^2} + 67.5 \, \frac{\text{lbf}}{\text{in}^2}$$

$$+ \frac{\left(43.8 \, \frac{\text{lbf}}{\text{in}^2} + 5 \, \frac{\text{lbf}}{\text{in}^2}\right)\left(28.5 \, \frac{\text{lbf}}{\text{in}^2} + 5 \, \frac{\text{lbf}}{\text{in}^2}\right)}{\left(43.8 \, \frac{\text{lbf}}{\text{in}^2} + 5 \, \frac{\text{lbf}}{\text{in}^2}\right) + \left(28.5 \, \frac{\text{lbf}}{\text{in}^2} + 5 \, \frac{\text{lbf}}{\text{in}^2}\right)}$$

$$= 110.6 \text{ psi}$$

The total head is

$$h_A = 95 \text{ ft} - 0 \text{ ft}$$
$$+ \frac{\left(110.6 \ \frac{\text{lbf}}{\text{in}^2}\right)\left(32.2 \ \frac{\text{ft-lbm}}{\text{lbf-sec}^2}\right)\left(12 \ \frac{\text{in}}{\text{ft}}\right)^2}{\left(62.4 \ \frac{\text{lbm}}{\text{ft}^3}\right)\left(32.2 \ \frac{\text{ft}}{\text{sec}^2}\right)}$$
$$= 350.2 \text{ ft}$$

The brake horsepower is

$$\text{BHP} = \frac{(350.1 \text{ ft})\left(350 \ \frac{\text{gal}}{\text{min}} + 860 \ \frac{\text{gal}}{\text{min}}\right)}{\left(3956 \ \frac{\text{ft-gal}}{\text{min-hp}}\right)(0.95)}$$
$$= 112.76 \text{ hp} \quad (113 \text{ hp})$$

**The answer is (D).**

Why Other Options Are Wrong

(A) This incorrect answer results when the pressure is not converted to lbf/ft$^2$.

(B) This incorrect answer results when the hydraulic power is multiplied by instead of divided by the efficiency.

(C) This answer results when the pump location height is erroneously considered as the suction height.

## SOLUTION 50

When two impellers of different size operate at the same speed and efficiency, the affinity laws state that

$$\frac{Q_2}{Q_1} = \frac{d_2}{d_1}$$
$$\frac{h_2}{h_1} = \left(\frac{d_2}{d_1}\right)^2$$
$$\frac{P_2}{P_1} = \left(\frac{d_2}{d_1}\right)^3$$

(These laws are only approximate in the case of a clipped impeller, but they are conventionally treated as exact because the difference is usually trivial.)

Thus,

$$Q_2 = Q_1\left(\frac{d_2}{d_1}\right)$$
$$= \left(200 \ \frac{\text{gal}}{\text{min}}\right)\left(\frac{1.5 \text{ ft}}{2.0 \text{ ft}}\right)$$
$$= 150 \text{ gpm}$$

$$h_2 = h_1\left(\frac{d_2}{d_1}\right)^2$$
$$= (180 \text{ ft})\left(\frac{1.5 \text{ ft}}{2.0 \text{ ft}}\right)^2 = 101 \text{ ft}$$

$$P_2 = P_1\left(\frac{d_2}{d_1}\right)^3$$
$$= (17.0 \text{ hp})\left(\frac{1.5 \text{ ft}}{2.0 \text{ ft}}\right)^3 = 7.2 \text{ hp}$$

**The answer is (D).**

Why Other Options Are Wrong

(A) This incorrect answer results when all three ratios are incorrectly thought to equal $d_2/d_1$.

(B) This answer results when the affinity laws are incorrectly stated as

$$\frac{Q_2}{Q_1} = 1$$
$$\frac{h_2}{h_1} = \frac{d_2}{d_1}$$
$$\frac{P_2}{P_1} = \left(\frac{d_2}{d_1}\right)^2$$

(C) This answer results when all three ratios are incorrectly thought to equal $(d_2/d_1)^3$.

## SOLUTION 51

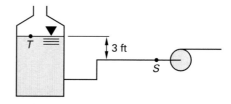

To prevent cavitation, the required NPSH must be less than or equal to the available NPSH, where

$$\text{NPSHA} = (z_T - z_S) + \frac{v_T^2}{2g} + \frac{p_T - p_{\text{sat}}}{\gamma} - h_{f,S}$$

Solving for $h_{f,S}$ gives

$$h_{f,S} < (z_T - z_S) + \frac{\mathrm{v}_T^2}{2g} + \frac{p_T - p_{\text{sat}}}{\gamma} - \text{NPSHR}$$

The velocity in the tank is 0 ft/sec, the water level in the tank is 3 ft higher than at the pump intake, atmospheric pressure is 14.7 psia, the saturation pressure of water at 180°F (from steam tables) is 7.52 psia, and the specific weight of water at 180°F is 60.6 lbm/ft$^3$.

$$h_{f,S} < (3 \text{ ft} - 0 \text{ ft}) + 0 \text{ ft}$$
$$+ \frac{\left(14.7 \frac{\text{lbf}}{\text{in}^2} - 7.52 \frac{\text{lbf}}{\text{in}^2}\right)\left(12 \frac{\text{in}}{\text{ft}}\right)^2}{60.6 \frac{\text{lbf}}{\text{ft}^3}} - 5.0 \text{ ft}$$
$$= 15.06 \text{ ft}$$

Solving for $L$ in the manufacturer's equation,

$$h_{f,S} = 1.50 \times 10^{-5} L Q^{1.852}$$

$$L < \frac{h_{f,S}}{(1.50 \times 10^{-5}) Q^{1.852}}$$

$$= \frac{15.06 \text{ ft}}{(1.50 \times 10^{-5})\left(140 \frac{\text{gal}}{\text{min}}\right)^{1.852}}$$

$$= 106 \text{ ft} \quad (100 \text{ ft})$$

**The answer is (B).**

Why Other Options Are Wrong

(A) This incorrect answer results when $z_T - z_S$ is left out of the calculations.

(C) This incorrect answer results when the subtraction of NPSHR is neglected.

(D) This incorrect answer results when $p_{\text{sat}}$ is left out of the calculations.

## SOLUTION 52

First convert the required flow of 23 in$^3$/sec to gallons per minute.

$$Q_{\text{req}} = \left(23 \frac{\text{in}^3}{\text{sec}}\right)\left(60 \frac{\text{sec}}{\text{min}}\right)\left(\frac{1 \text{ gal}}{231 \text{ in}^3}\right)$$
$$= 5.97 \text{ gal/min}$$

Begin by testing the 5 gpm valve because it is one of the middle answers. Calculating the constant $KI_m$ using the manufacturer's equation and pressure drop value gives

$$KI_m = \frac{Q_R}{\sqrt{P_v}} = \frac{5 \frac{\text{gal}}{\text{min}}}{\sqrt{1000 \frac{\text{lbf}}{\text{in}^2}}} = 0.158 \text{ gal-in/min-lbf}^{1/2}$$

Now the equation can be used again to determine what the flow rate through the 5 gpm valve would be for the actual pressure drop.

$$Q_{5\,\text{gpm}} = KI_m \sqrt{P_v}$$
$$= \left(0.158 \frac{\text{gal-in}}{\text{min-lbf}^{1/2}}\right)\sqrt{1500 \frac{\text{lbf}}{\text{in}^2} - 1100 \frac{\text{lbf}}{\text{in}^2}}$$
$$= 3.16 \text{ gal/min}$$

The flow rate that the 5 gpm valve would give is less than the required flow rate, so this valve is not large enough. The quickest way to finish this problem from this point is to notice that $Q$ is linear with respect to $KI_m$ and that $KI_m$ is linear with respect to $Q_R$. Thus, doubling $Q_R$ to 10 gpm will double $Q$ to 6.32 gpm, which satisfies the flow rate requirement.

Double-check the capacity of the 10 gpm valve. The constant for this valve is

$$KI_m = \frac{10 \frac{\text{gal}}{\text{min}}}{\sqrt{1000 \frac{\text{lbf}}{\text{in}^2}}} = 0.316 \text{ gal-in/min-lbf}^{1/2}$$

The flow rate through this valve under actual conditions is

$$Q_{10\,\text{gpm}} = \left(0.316 \frac{\text{gal-in}}{\text{min-lbf}^{1/2}}\right)$$
$$\times \sqrt{1500 \frac{\text{lbf}}{\text{in}^2} - 1100 \frac{\text{lbf}}{\text{in}^2}}$$
$$= 6.32 \text{ gal/min}$$

This is greater than the required flow rate, so the 10 gpm valve will suffice. The 15 gpm valve would also provide a large enough flow rate, but it is probably more expensive than the 10 gpm valve. So the best choice is the 10 gpm valve.

**The answer is (C).**

# DEPTH SOLUTIONS 49

## Why Other Options Are Wrong

(A) This incorrect solution results from not converting seconds to minutes in the calculation of the required flow rate.

(B) This incorrect solution results from overlooking that the flow rate of 5 gpm quoted by the manufacturer is based on a pressure drop of 1000 psi. In this particular application the pressure drop is only 400 psi, so the actual flow rate through the valve would be significantly less than 5 gpm.

(D) This valve has a large enough flow rate, but it is not the least expensive valve that would work in the application.

## SOLUTION 53

Item I is a pipe fitting classification and is an abbreviation for American National Standard taper pipe thread. NPT is also used as an abbreviation for the following.

American Standard Pipe Tapered Thread
National Pipe Thread
National Pipe Tapered

Item II is a pipe fitting classification and is an abbreviation for the Society of Automotive Engineers.

Item III is a screw thread classification and is an abbreviation for Unified Coarse.

Item IV is the abbreviation for the American Society of Mechanical Engineers.

**The answer is (A).**

## Why Other Options Are Wrong

(B) This answer is incorrect because UNC is a screw thread classification, not a pipe fitting classification.

(C) This answer is incorrect because UNC is a screw thread classification, not a pipe fitting classification.

(D) This answer is incorrect because neither UNC nor ASME is a pipe fitting classification.

## SOLUTION 54

The throttling area in a needle valve is a function of the seat opening diameter, needle lift, and cone half angle.

**The answer is (C).**

## Why Other Options Are Wrong

(A) This answer is incorrect because throttling area is a function of needle lift.

(B) This answer is incorrect because throttling area is a function of seat opening diameter.

(D) This answer is incorrect because throttling area is a function of needle lift, and not a function of cone height.

## SOLUTION 55

The lohm law equation is

$$L = \frac{20}{Q}\sqrt{\frac{H}{SG}}$$

The flow rate is $Q$ (gpm), the differential pressure is $H$ (psi), and the specific gravity is SG.

Using the given values, the flow resistance is found to be

$$L = \left(\frac{20\,\frac{\text{gal}}{\text{min-in-lbf}^{1/2}}}{1.5\,\frac{\text{gal}}{\text{min}}}\right)\sqrt{\frac{1000\,\frac{\text{lbf}}{\text{in}^2}}{1.09}}$$

$$= 404\text{ in}^{-2} \quad (400 \text{ lohms})$$

**The answer is (A).**

## Why Other Options Are Wrong

(B) This answer is incorrect because the flow rate was multiplied by the constant 20 instead of being divided into it.

(C) This answer is incorrect because the flow rate $I$ was converted to cubic feet per minute instead of gallons per minute.

(D) This answer is incorrect because the square root of the $H/\text{SG}$ ratio was not taken.

## SOLUTION 56

The power generated by the turbine can be found using

$$P = \eta_t \dot{m}\Delta h = \eta_t \dot{m}(h_{\text{inlet}} - h_{\text{exhaust}})$$

The turbine efficiency is $\eta_t$. Assuming the expansion is isentropic, the entropy at the turbine inlet and exhaust will be the same. This will be used to determine the enthalpy at the exhaust. Using steam tables at 1300 psia and 940°F,

$$h_{\text{inlet}} = 1460.8 \text{ Btu/lbm}$$
$$s_{\text{inlet}} = 1.5941 \text{ Btu/lbm-°F}$$

Assuming isentropic expansion, $s_{\text{exhaust}}$ is equal to 1.5941 Btu/lbm-°F. Using this value and a pressure of 300 psia, the steam tables give an enthalpy of

$$h_{\text{exhaust}} = 1281 \text{ Btu/lbm}$$

PROFESSIONAL PUBLICATIONS, INC.

No interpolation is needed since the entropy is 1.5940 Btu/lbm-°F, which is close enough to $s_{\text{inlet}}$.

Substituting into the power equation,

$$9 \times 10^6 \text{ W} = 0.75 \dot{m} \left(1460.8 \frac{\text{Btu}}{\text{lbm}} - 1281 \frac{\text{Btu}}{\text{lbm}}\right)$$

$$\times \left(\frac{1 \text{ W}}{3.413 \frac{\text{Btu}}{\text{hr}}}\right)$$

$$\dot{m} = 228 \times 10^3 \text{ lbm/hr}$$

**The answer is (D).**

**Why Other Options Are Wrong**

(A) This incorrect answer results when the conversion from watts to Btu/hr is neglected.

(B) This incorrect answer results when the internal energy of the superheated steam is used instead of the enthalpy at 1300 psia and 940°F.

(C) This incorrect answer results when the efficiency of the system is ignored.

## SOLUTION 57

The NPSHA is defined as

$$\text{NPSHA} = h_{\text{atm}} + h_{z,S} - h_{f,S} - h_{\text{vp}}$$

Item I will increase the NPSHA because a larger diameter pipe will reduce the fluid velocity. This will reduce the friction loss, $h_f$.

Item II will *not* increase the NPSHA because throttling the input will increase the fluid velocity. This will increase the friction head loss and decrease the NPSHA.

Item III will increase the NPSHA because pressurizing the supply tank will increase the atmospheric head, $h_a$.

Item IV will increase the NPSHA because increasing the height increases the static suction head, $h_s$.

**The answer is (D).**

**Why Other Options Are Wrong**

(A) This answer is wrong because item III will also increase the NPSHA.

(B) This answer is wrong because item II will not increase the NPSHA.

(C) This answer is wrong because item II will not increase the NPSHA.

## SOLUTION 58

The mechanical runout is the largest difference between any two readings. The smallest reading is $-0.1$ mil and the largest reading is $0.15$ mil. The difference is $0.15 \text{ mil} - (-0.1 \text{ mil}) = 0.25$ mil.

**The answer is (D).**

**Why Other Options Are Wrong**

(A) This incorrect answer is the average of the dial indicator readings, not the runout of the shaft.

(B) This incorrect answer is the largest negative value from zero, not the runout of the shaft.

(C) This incorrect answer is the largest positive value from zero, not the runout of the shaft.

## SOLUTION 59

The pump efficiency can be estimated by comparing the motor's power output to the pump's power consumption.

Starting with the motor efficiency, the power output by the motor is

$$\eta_m = \frac{P_{m,\text{out}}}{P_{m,\text{in}}}$$

$$P_{m,\text{out}} = \eta_m P_{m,\text{in}} = \eta_m \sqrt{3} V I \cos\phi$$

The motor efficiency and power factor are $\eta_m$ and $\cos\phi$, respectively.

$$P_{m,\text{out}} = (0.93)\sqrt{3}(2300 \text{ V})(100 \text{ A})(0.87)\left(\frac{1 \text{ kW}}{1000 \text{ W}}\right)$$

$$= 322 \text{ kW}$$

Using the pump efficiency, the power needed by the pump, $P_{p,\text{in}}$, in terms of the differential pressure is

$$\eta_p = \frac{P_{p,\text{out}}}{P_{p,\text{in}}}$$

$$\eta_p P_{p,\text{in}} = P_{p,\text{out}} = Q\Delta p$$

$$= \left(500 \frac{\text{gal}}{\text{min}}\right)\left(\frac{1 \text{ ft}^3}{7.48 \text{ gal}}\right)\left(\frac{1 \text{ min}}{60 \text{ sec}}\right)\left(850 \frac{\text{lbf}}{\text{in}^2}\right)$$

$$\times \left(12 \frac{\text{in}}{\text{ft}}\right)^2 \left(1.356 \times 10^{-3} \frac{\frac{\text{kW}}{\text{ft-lbf}}}{\text{sec}}\right)$$

$$P_{p,\text{in}} = \frac{185 \text{ kW}}{\eta_p}$$

By equating the output power from the motor with the input power needed by the pump, an approximate pump efficiency can be obtained.

$$322 \text{ kW} = \frac{185 \text{ kW}}{\eta_p}$$

$$\eta_p = 0.57 \quad (57\%)$$

**The answer is (A).**

Why Other Options Are Wrong

(B) This incorrect answer results from neglecting to make the conversion from horsepower to kilowatts.

(C) This incorrect answer results from neglecting to use the power factor and the square root of three.

(D) This incorrect answer results from neglecting to use the square root of three.

## SOLUTION 60

The power can be estimated using

$$P = \left(\frac{\dot{m}\omega}{g_c}\right)(v_{t,\text{out}} r_o - v_{t,\text{in}} r_i)$$

The mass flow rate, $\dot{m}$, angular velocity, $\omega$, tangential velocity of the fluid leaving the impeller, $v_{t,\text{out}}$, and the outer radius of the impeller, $r_o$, are given. The tangential velocity of the fluid entering the impeller, $v_{t,\text{in}}$, is usually negligible, so the power equation becomes

$$P = \left(\frac{\dot{m}\omega}{g_c}\right) v_{t,\text{out}} r_o$$

Substituting given values,

$$P = \left(\frac{\left(900 \frac{\text{lbm}}{\text{sec}}\right)\left(6300 \frac{\text{rev}}{\text{min}}\right)\left(\frac{1 \text{ min}}{60 \text{ sec}}\right)\left(2\pi \frac{\text{rad}}{\text{rev}}\right)}{32.2 \frac{\text{ft-lbm}}{\text{lbf-sec}^2}}\right)$$

$$\times \left(300 \frac{\text{ft}}{\text{sec}}\right)(12 \text{ in})\left(\frac{1 \text{ ft}}{12 \text{ in}}\right)\left(\frac{1 \text{ hp}}{550 \frac{\text{ft-lbf}}{\text{sec}}}\right)$$

$$= 10{,}000 \text{ hp}$$

**The answer is (D).**

Why Other Options Are Wrong

(A) This incorrect answer results from dividing by 3600 instead of 60 when converting to rad/sec.

(B) This incorrect answer results from neglecting the factor of $2\pi$ in the conversion from rpm to rad/sec.

(C) This incorrect answer results from transposing the 6 and the 3 in the rpm value.

## SOLUTION 61

As the flow rate decreases from point A, the compressor performance moves up the pressure-ratio/flow-rate curve to point B. Further reduction in the flow rate causes the compressor to operate at point C, reducing the compressor's pressure capability. As the compressor's performance moves to point C, refrigerant (in the case of refrigeration operation) continues to boil off because of the evaporator's heat load. This builds up the evaporator pressure and decreases the pressure ratio, causing the compressor to shift briefly back to point A where the cycle repeats. This cycle is known as surging and typically occurs when the compressor is operating at less than 35% of the rated capacity.

**The answer is (A).**

Why Other Options Are Wrong

(B) Stalling is the breakdown of the airflow through a few stages of a compressor.

(C) Starving is caused by lack of air at the inlet of the compressor.

(D) Throttling is the regulation of flow and head.

## SOLUTION 62

The flow rate needed to achieve a pump efficiency of 75% can be found using an efficiency versus specific speed graph, such as is found in the *Mechanical Engineering Reference Manual*. In this graph the 75% efficiency line intersects both the 200 gal/min and 500 gal/min curves. To achieve 75% efficiency at 200 gal/min requires a specific speed of 2000, and at 500 gal/min a specific speed of 1000 is required.

The specific speed is given by the formula

$$n_s = \frac{n\sqrt{Q}}{h_A^{0.75}}$$

For a flow rate of 200 gal/min, the specific speed is

$$n_s = \frac{\left(3550 \frac{\text{rev}}{\text{min}}\right)\sqrt{200 \frac{\text{gal}}{\text{min}}}}{(340 \text{ ft})^{0.75}} = 634$$

Specific speed is not dimensionless, but the units are meaningless, so specific speed is expressed as a unitless number.

At this flow rate, the specific speed is 634, not the required 2000. Therefore, this is not the correct flow rate. In fact, at this specific speed and flow rate the pump efficiency is approximately 60%.

For a flow rate of 500 gal/min, the specific speed is

$$n_s = \frac{\left(3550 \, \frac{\text{rev}}{\text{min}}\right)\sqrt{500 \, \frac{\text{gal}}{\text{min}}}}{(340 \text{ ft})^{0.75}} = 1000$$

This flow rate provides the needed specific speed to achieve a pump efficiency of 75%.

**The answer is (B).**

Why Other Options Are Wrong

(A) This incorrect flow rate only provides a pump efficiency of 60% as shown.

(C) This flow rate was derived using an incorrect total head of 430 ft.

(D) This answer is incorrect because a pump efficiency of 75% can be achieved.

## SOLUTION 63

The suction lift can be calculated using the net positive suction head available (NPSHA) equation

$$\text{NPSHA} = h_{\text{atm}} + h_{z,S} + h_{f,S} - h_{\text{vp}}$$

From a table of water properties, the specific weight of water at 80°F is 62.2 lbf/ft³. Therefore, the atmospheric head is

$$h_{\text{atm}} = \frac{p}{\gamma} = \frac{\left(14.7 \, \frac{\text{lbf}}{\text{in}^2}\right)\left(12 \, \frac{\text{in}}{\text{ft}}\right)^2}{62.2 \, \frac{\text{lbf}}{\text{ft}^3}}$$

$$= 34 \text{ ft}$$

Also from a table of water properties, the vapor pressure head of water $h_{\text{vp}}$ at 80°F is 1.17 ft.

The friction loss head is 5 ft. The static suction head can now be calculated.

$$\text{NPSHA} = h_{\text{atm}} + h_{z,S} + h_{f,S} - h_{\text{vp}}$$
$$10 \text{ ft} = 34 \text{ ft} + h_{z,S} - 5 \text{ ft} - 1.17 \text{ ft}$$
$$h_{z,S} = -17.8 \text{ ft} \quad (-18 \text{ ft})$$

The negative sign indicates that the pump is above the free level of the supply surface and there is 18 ft of suction lift.

**The answer is (B).**

Why Other Options Are Wrong

(A) This answer results when the net positive suction head and the suction lift are incorrectly thought to be equal.

(C) This incorrect answer results when the atmospheric head is taken as 43 ft instead of 34 ft.

(D) This incorrect answer results when the friction loss head and vapor pressure head are added to the right side of the NPSHA equation instead of subtracted.

## SOLUTION 64

The heat transfer rate needed to obtain the required steam production is

$$Q = \dot{m}\Delta h_{\text{water}}$$
$$= \dot{m}(h_{\text{steam}} - h_{\text{water}})$$

The enthalpy of the saturated water entering the drum boiler at 600°F is

$$h_{\text{water}} = 616.7 \, \frac{\text{Btu}}{\text{lbm}}$$

The enthalpy of steam at 1900 psia and 95% quality can be determined from the saturated steam tables and the definition of the quality.

$$h_{\text{steam}} = h_f + x h_{fg}$$
$$= 660.1 \, \frac{\text{Btu}}{\text{lbm}} + (0.95)\left(483.4 \, \frac{\text{Btu}}{\text{lbm}}\right)$$
$$= 1119.3 \text{ Btu/lbm}$$

Only 92% of the transferred heat is used, so the total heat transferred from the flue gas must be

$$Q_{\text{total}} = \frac{\dot{m}(h_{\text{steam}} - h_{\text{water}})}{\eta}$$
$$= \frac{\left(540{,}000 \, \frac{\text{lbm}}{\text{hr}}\right)\left(1119.3 \, \frac{\text{Btu}}{\text{lbm}} - 616.7 \, \frac{\text{Btu}}{\text{lbm}}\right)}{0.92}$$
$$= 295{,}000{,}000 \text{ Btu/hr}$$

The heat transfer rate, as a function of the temperature difference, is given by

$$Q = UA(T_1 - T_2)$$

Solving for the heat-exchange area,

$$A = \frac{Q}{U(T_1 - T_2)}$$

$$= \frac{295{,}000{,}000 \ \dfrac{\text{Btu}}{\text{hr}}}{\left(218.2 \ \dfrac{\text{Btu}}{\text{hr-ft}^2\text{-}°\text{F}}\right)(32°\text{F})}$$

$$= 42{,}249 \ \text{ft}^2$$

The total number of tubes required is

$$n_{\text{tubes}} = \frac{A}{A_{\text{tube}}}$$

$$= \frac{42{,}249 \ \text{ft}^2}{5.3 \ \dfrac{\text{ft}^2}{\text{tube}}}$$

$$= 7972 \ \text{tubes} \quad (8000 \ \text{tubes})$$

**The answer is (B).**

Why Other Options Are Wrong

(A) This incorrect answer results when the total heat transfer is multiplied by the efficiency instead of divided by it.

(C) This incorrect answer results when the steam quality is omitted when calculating the steam enthalpy.

(D) This incorrect answer results when the temperature difference is neglected in the calculation of the required area.

## SOLUTION 65

The actual cubic feet per minute, ACFM, is found using the flow rate at standard conditions known as standard cubic feet per minute, SCFM, and the following equation.

$$\text{ACFM} = K_d(\text{SCFM})$$

The constant $K_d$ is a density correction factor for altitude and is given by

$$K_d = \frac{\rho_{\text{std}}}{\rho_{\text{actual}}} = \left(\frac{p_{\text{std}}}{p_{\text{actual}}}\right)\left(\frac{T_{\text{actual}}}{T_{\text{std}}}\right)$$

Most tables give the ratio $p_{\text{actual}}/p_{\text{std}}$. At 6000 ft this ratio is

$$\frac{p_{\text{actual}}}{p_{\text{std}}} = 0.801$$

Converting the air temperature to degrees Rankine, the density correction factor is therefore

$$K_d = (0.801)^{-1}\left(\frac{70°\text{F} + 460°}{70°\text{F} + 460°}\right) = \frac{1}{0.801}$$

The actual flow rate of the fan is

$$\text{ACFM} = K_d(\text{SCFM}) = \left(\frac{1}{0.801}\right)\left(4700 \ \frac{\text{ft}^3}{\text{min}}\right)$$

$$= 5870 \ \text{ft}^3/\text{min} \quad (5900 \ \text{ft}^3/\text{min})$$

**The answer is (B).**

Why Other Options Are Wrong

(A) This incorrect answer results from using the ratio of $\rho_{\text{actual}}/\rho_{\text{std}}$ instead of its reciprocal.

(C) This incorrect answer results from reading an incorrect $K_d$ value of 0.772 from the table.

(D) This incorrect answer results from transposing 4 and 7 in the flow rate.

## SOLUTION 66

The brake horsepower for the compressor can be determined using this graph, which gives the brake horsepower needed for compressing natural gas.

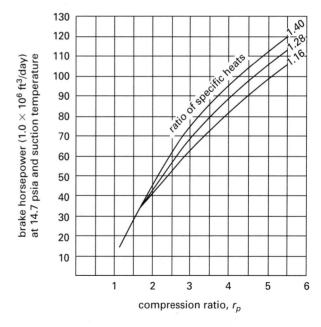

Reprinted from *Rules of Thumb for Mechanical Engineers*, Pope ed., p. 118, Copyright 1996, with permission from Elsevier.

The compression ratio, $r_p$, is calculated using the formula

$$r_p = \frac{p_d}{p_s}$$

The discharge pressure, $p_d$, and the suction pressure, $p_s$, must be absolute pressures. The atmospheric pressure at 3000 ft is 13.2 psia, so the compression ratio is

$$r_p = \frac{200 \dfrac{\text{lbf}}{\text{in}^2} + 13.2 \dfrac{\text{lbf}}{\text{in}^2}}{35 \dfrac{\text{lbf}}{\text{in}^2} + 13.2 \dfrac{\text{lbf}}{\text{in}^2}} = 4.42$$

From the graph, the brake horsepower required for $1.0 \times 10^6$ ft³/day is approximately 95 hp. For $1.7 \times 10^6$ ft³/day, then, the horsepower requirement is

$$P = (95 \text{ hp}) \left( \frac{1.7 \times 10^6 \dfrac{\text{ft}^3}{\text{day}}}{1.0 \times 10^6 \dfrac{\text{ft}^3}{\text{day}}} \right)$$

$$= 162 \text{ hp} \quad (160 \text{ hp})$$

**The answer is (C).**

Why Other Options Are Wrong

(A) This incorrect answer results when the correction factor for $1.7 \times 10^6$ ft³/day is neglected.

(B) This incorrect answer results when absolute pressures are not used when the compression ratio is calculated, and the correction for $1.7 \times 10^6$ ft³/day is neglected.

(D) This incorrect answer results when absolute pressures are not used when the compression ratio is calculated.

## SOLUTION 67

Because the compressor is not cooled, the power necessary for isentropic compression of an ideal gas from $p_1$ to $p_2$ is known as the theoretical adiabatic power and is given by

$$P_{\text{th}} = \left( \frac{k}{k-1} \right) Q_1 p_1 \left( \left( \frac{p_2}{p_1} \right)^{\frac{k-1}{k}} - 1 \right)$$

The ratio of specific heats is $k$, the flow rate is $Q_1$, the inlet pressure is $p_1$, and the discharge pressure is $p_2$. The ratio of specific heats is 1.4 for air, and the inlet pressure for the atmospheric air that enters this compressor is 14.7 psia. The adiabatic power is

$$P_{\text{th}} = \left( \frac{1.4}{1.4-1} \right) \left( 50{,}000 \frac{\text{ft}^3}{\text{min}} \right) \left( \frac{1 \text{ min}}{60 \text{ sec}} \right) \left( 14.7 \frac{\text{lbf}}{\text{in}^2} \right)$$

$$\times \left( 144 \frac{\text{in}^2}{\text{ft}^2} \right) \left( \left( \frac{53 \dfrac{\text{lbf}}{\text{in}^2}}{14.7 \dfrac{\text{lbf}}{\text{in}^2}} \right)^{\frac{1.4-1}{1.4}} - 1 \right)$$

$$\times \left( \frac{1 \text{ hp}}{550 \dfrac{\text{lbf-ft}}{\text{sec}}} \right)$$

$$= 4968 \text{ hp}$$

The compressor is 79% efficient, so the motor shaft horsepower must be

$$P_{\text{shaft}} = \frac{P_{\text{th}}}{\eta}$$

$$= \frac{4968 \text{ hp}}{0.79}$$

$$= 6289 \text{ hp}$$

**The answer is (D).**

Why Other Options Are Wrong

(A) This incorrect answer results when the flow is converted from cfm to cfs by dividing by $(60)^2$ instead of 60.

(B) This incorrect answer results when the specific heats term at the beginning of the adiabatic power equation is ignored.

(C) This incorrect answer results when 35 psia is used as the discharge pressure instead of 53 psia.

## SOLUTION 68

This problem can be solved with the correlation equation that uses the actual water loading, $L_a$, the measured air loading, $G_a$, the water loading at test temperatures, $L_b$, and the base air loading, $G_b$.

$$\frac{L_b}{L_a} = \left( \frac{G_b}{G_a} \right)^n$$

At constant conditions and using an average value of $n = 0.93$ the above equation becomes

$$\frac{G_b}{G_a} = \left( \frac{L_b}{L_a} \right)^{1/n} = \left( \frac{608 \text{ ft}^2}{576 \text{ ft}^2} \right)^{1/0.93} = 1.06$$

Therefore the air rate must be increased by 6%.

**The answer is (C).**

Why Other Options Are Wrong

(A) This incorrect answer results when 0.39 is used as the exponent instead of $1/0.93$.

(B) This incorrect answer results when 0.93 is used as the exponent instead of $1/0.93$.

(D) This incorrect answer results when $1/0.39$ is used as the exponent instead of $1/0.93$.

## SOLUTION 69

A boiler horsepower is defined as approximately 33,475 Btu/hr, or 9.81 kW, and is the amount of power needed to convert 34.5 lbm/hr of feedwater at 212°F and atmospheric pressure to dry, saturated steam at the same temperature and pressure. Answer IV can be seen to be true from the following calculation.

From the steam tables at 212°F,

$$h_{fg} = 970.3 \text{ Btu/lbm}$$

$$P = \dot{m}\Delta h = \left(34.5 \ \frac{\text{lbm}}{\text{hr}}\right)\left(970.3 \ \frac{\text{Btu}}{\text{lbm}}\right)$$

$$= 33{,}475 \text{ Btu/hr}$$

**The answer is (C).**

Why Other Options Are Wrong

(A) This answer is incorrect because IV is also true.

(B) This answer is incorrect because III is not true, and IV is.

(D) This answer is incorrect because II and III are both false.

# SYSTEMS

## SOLUTION 70

Method 1

A general rule of thumb for quickly calculating the inside pipe diameter is

$$d_i = \sqrt{\frac{Q}{20}} = \sqrt{\frac{1750 \ \frac{\text{gal}}{\text{min}}}{20 \ \frac{\text{gal}}{\text{min-in}^2}}} = 9.35 \text{ in}$$

The smallest available pipe, then, is 12 in.

Method 2

The recommended maximum velocity for water flowing through a city's water pipeline is 2 to 5 ft/sec. Using the flow rate equation, the inner pipe diameter can be found. Using the largest velocity in the flow rate equation, the smallest inner pipe diameter can be found.

$$Q = vA$$

$$\left(1750 \ \frac{\text{gal}}{\text{min}}\right)\left(\frac{1 \text{ ft}^3}{7.48 \text{ gal}}\right)\left(\frac{1 \text{ min}}{60 \text{ sec}}\right)\left(12 \ \frac{\text{in}}{\text{ft}}\right)^3$$

$$= \left(5 \ \frac{\text{ft}}{\text{sec}}\right)\left(12 \ \frac{\text{in}}{\text{ft}}\right)\left(\frac{\pi}{4}d^2\right)$$

$$d = 11.9 \text{ in}$$

The smallest schedule-40 pipe with the required inner diameter is 12 in.

**The answer is (B).**

Why Other Options Are Wrong

(A) This incorrect solution results from dividing by 20 ft$^3$/min-in$^2$ instead of 20 gal/min-in$^2$.

(C) This incorrect solution results from dividing by 2 gal/min-in$^2$ instead of 20 gal/min-in$^2$.

(D) This incorrect solution results from dividing by 20 ft$^3$/min-ft$^2$ instead of 20 gal/min-in$^2$.

## SOLUTION 71

Method 1

The following equation can be used to approximate the time needed to increase the temperature of a boiler from $T_1$ to $T_2$.

$$\ln\left(\frac{T_{g,\text{in}} - T_1}{T_{g,\text{in}} - T_2}\right) = \left(\frac{\dot{m}_g c_{p,g}\left(e^{\frac{UA}{\dot{m}_g c_p}} - 1\right)}{Ce^{\frac{UA}{\dot{m}_g c_p}}}\right)\Delta t$$

The entering flue gas temperature, $T_{g,\text{in}}$, is 1300°F; the mass flow rate of the flue gases, $\dot{m}_g$, is 100,000 lbm/hr; the flue gas specific heat, $c_{p,g}$, is 0.25 Btu/lbm-°F; the overall heat transfer coefficient, $U$, is 5 Btu/ft$^2$-hr-°F; and the surface area of the boiler, $A$, is 30,000 ft$^2$.

The thermal capacitance of the boiler, $C$, must be calculated as follows.

$$C = m_{\text{steel}} c_{p,\text{steel}} + m_{\text{water}} c_{p,\text{water}}$$

$$= (70{,}000 \text{ lbm})\left(0.11 \ \frac{\text{Btu}}{\text{lbm-°F}}\right)$$

$$+ (45{,}000 \text{ lbm})\left(1.0 \ \frac{\text{Btu}}{\text{lbm-°F}}\right)$$

$$= 52{,}700 \text{ Btu/°F}$$

The exponent of the two exponential terms is

$$\frac{UA}{\dot{m}_g c_p} = \frac{\left(5 \frac{\text{Btu}}{\text{ft}^2\text{-hr-}°\text{F}}\right)(30{,}000 \text{ ft}^2)}{\left(100{,}000 \frac{\text{lbm}}{\text{hr}}\right)\left(0.25 \frac{\text{Btu}}{\text{lbm-}°\text{F}}\right)} = 6$$

Using these equations and the given values, the time needed to heat the boiler can be calculated.

$$\ln\left(\frac{1300°\text{F} - 90°\text{F}}{1300°\text{F} - 212°\text{F}}\right) = \left(\frac{\left(100{,}000 \frac{\text{lbm}}{\text{hr}}\right)\left(\frac{1 \text{ hr}}{60 \text{ min}}\right) \times \left(0.25 \frac{\text{Btu}}{\text{lbm-}°\text{F}}\right)(e^6 - 1)}{\left(52{,}700 \frac{\text{Btu}}{°\text{F}}\right)e^6}\right)\Delta t$$

$$\Delta t = 13.5 \text{ min} \quad (14 \text{ min})$$

**The answer is (B).**

## Method 2

The solution can also be obtained through the fundamental concepts used to derive the equation used in Method 1. First, the energy balance of the system can be stated this way: The change in internal energy of the boiler during $dt$ is equal to the net heat flow from the flue gases to the boiler during $dt$. If $C$ is the thermal capacitance of the boiler, then the energy balance equation is

$$C(dT) = \dot{m}_g c_{p,g}(T_{g,\text{in}} - T_{g,\text{out}})dt$$

The heat flow rate to the surface of the boiler can be expressed using the overall heat transfer coefficient and the logarithmic mean temperature difference (LMTD), $\Delta \overline{T}$, as

$$Q = UA\Delta\overline{T}$$

If $T$ is the temperature of the water in the boiler, the LMTD is

$$\Delta\overline{T} = \frac{(T_{g,\text{in}} - T) - (T_{g,\text{out}} - T)}{\ln\left(\frac{T_{g,\text{in}} - T}{T_{g,\text{out}} - T}\right)}$$

$$= \frac{(1300°\text{F} - T) - (T_{g,\text{out}} - T)}{\ln\left(\frac{1300°\text{F} - T}{T_{g,\text{out}} - T}\right)}$$

Equating this heat flow rate with the heat flow term on the right side of the energy balance equation gives

$$UA\left(\frac{(1300°\text{F} - T) - (T_{g,\text{out}} - T)}{\ln\left(\frac{1300°\text{F} - T}{T_{g,\text{out}} - T}\right)}\right)$$

$$= \dot{m}_g c_{p,g}(T_{g,\text{in}} - T_{g,\text{out}})$$

Substituting known values,

$$\left(5 \frac{\text{Btu}}{\text{ft}^2\text{-hr-}°\text{F}}\right)(30{,}000 \text{ ft}^2)$$

$$\times \left(\frac{(1300°\text{F} - T) - (T_{g,\text{out}} - T)}{\ln\left(\frac{1300°\text{F} - T}{T_{g,\text{out}} - T}\right)}\right)$$

$$= \left(100{,}000 \frac{\text{lbm}}{\text{hr}}\right)\left(0.25 \frac{\text{Btu}}{\text{lbm-}°\text{F}}\right)(1300°\text{F} - T_{g,\text{out}})$$

Solving for $T_{g,\text{out}}$,

$$\left(150{,}000 \frac{\text{Btu}}{\text{hr-}°\text{F}}\right)(1300°\text{F} - T_{g,\text{out}})$$

$$= \left(25{,}000 \frac{\text{Btu}}{\text{hr-}°\text{F}}\right)(1300°\text{F} - T_{g,\text{out}})$$

$$\times \ln\left(\frac{1300°\text{F} - T}{T_{g,\text{out}} - T}\right)$$

$$6 = \ln\left(\frac{1300°\text{F} - T}{T_{g,\text{out}} - T}\right)$$

$$T_{g,\text{out}} = (1300°\text{F} - T)e^{-6} + T$$

Substituting this expression for $T_{g,\text{out}}$ into the energy balance equation gives

$$C(dT) = \dot{m}_g c_{p,g}\left(T_{g,\text{in}} - \left((1300°\text{F} - T)e^{-6} + T\right)\right)dt$$

Substituting known values,

$$C(dT) = \left(100{,}000 \frac{\text{lbm}}{\text{hr}}\right)\left(0.25 \frac{\text{Btu}}{\text{lbm-}°\text{F}}\right)$$

$$\times \left(1300°\text{F} - \left((1300°\text{F} - T)e^{-6} + T\right)\right)dt$$

Using the thermal capacitance found previously and simplifying,

$$\left(52{,}700 \frac{\text{Btu}}{°\text{F}}\right)dT = \left(25{,}000 \frac{\text{Btu}}{\text{hr-}°\text{F}}\right)$$

$$\times (1297°\text{F} - 0.9975T)dt$$

The differential equation can now be integrated to find the time needed to heat the boiler from 90°F to 212°F. Separating variables,

$$\left(\dfrac{52{,}700\ \frac{\text{Btu}}{°\text{F}}}{1297°\text{F} - 0.9975T}\right) dT = \left(25{,}000\ \dfrac{\text{Btu}}{\text{hr-}°\text{F}}\right) dt$$

Integrating,

$$\int_{90°\text{F}}^{212°\text{F}} \left(\dfrac{52{,}700\ \frac{\text{Btu}}{°\text{F}}}{1297°\text{F} - 0.9975T}\right) dT = \int_0^{t_f} \left(25{,}000\ \dfrac{\text{Btu}}{\text{hr-}°\text{F}}\right) dt$$

$$52{,}832\ \dfrac{\text{Btu}}{°\text{F}} \int_{90°\text{F}}^{212°\text{F}} \left(\dfrac{1}{1300°\text{F} - T}\right) = \left(25{,}000\ \dfrac{\text{Btu}}{\text{hr-}°\text{F}}\right) t_f$$

$$\left(52{,}832\ \dfrac{\text{Btu}}{°\text{F}}\right)$$
$$\times \left(-\ln(1300°\text{F} - T)\right)\Big|_{90°\text{F}}^{212°\text{F}} = \left(25{,}000\ \dfrac{\text{Btu}}{\text{hr-}°\text{F}}\right) t_f$$

$$\left(52{,}832\ \dfrac{\text{Btu}}{°\text{F}}\right)$$
$$\times \big(-\ln(1300°\text{F} - 212°\text{F})$$
$$+ \ln(1300°\text{F} - 90°\text{F})\big) = \left(25{,}000\ \dfrac{\text{Btu}}{\text{hr-}°\text{F}}\right) t_f$$

Solving for the time needed to heat the boiler,

$$t_f = (0.225\ \text{hr})\left(60\ \dfrac{\text{min}}{\text{hr}}\right) = 13.5\ \text{min}\quad (14\ \text{min})$$

**The answer is (B).**

Why Other Options Are Wrong

(A) This incorrect answer results when only the mass of the water is used to determine the thermal capacity.

(C) This incorrect answer results when $T_1$ is not subtracted from $T_{g,\text{in}}$ in the numerator.

(D) This incorrect answer results when the specific heat of steel is used instead of that of the flue gas.

## SOLUTION 72

The cavitation number, $\sigma$, is calculated using the equation

$$\sigma = \dfrac{2g_c(p - p_{\text{vp}})}{\rho v^2}$$

The pressure, $p$, and the vapor pressure, $p_{\text{vp}}$, are absolute pressures. For no cavitation to occur, the cavitation number, $\sigma$, must be greater than the critical cavitation number, $\sigma_{\text{cr}}$. The vapor pressure is the saturation pressure, so

$$p_{\text{vp}} = \text{water pressure} + \text{atmospheric pressure}$$
$$= 120\ \dfrac{\text{lbf}}{\text{in}^2} + 14.7\ \dfrac{\text{lbf}}{\text{in}^2}$$
$$= 134.7\ \text{psia}$$

To find the correct water density, the temperature of the water must be known. From the saturated steam tables, the temperature of saturated water at 134.7 psia is approximately 350°F. The water density at this temperature is 55.6 lbm/ft$^3$.

Using the critical cavitation number and the equation for $\sigma$, the maximum flow velocity with no cavitation can be found.

$$1.2 = \dfrac{(2)\left(32.2\ \dfrac{\text{ft-lbm}}{\text{lbf-sec}^2}\right) \times \left(200\ \dfrac{\text{lbf}}{\text{in}^2} + 14.7\ \dfrac{\text{lbf}}{\text{in}^2} - 134.7\ \dfrac{\text{lbf}}{\text{in}^2}\right) \times \left(12\ \dfrac{\text{in}}{\text{ft}}\right)^2}{\left(55.6\ \dfrac{\text{lbm}}{\text{ft}^3}\right) v^2}$$

$$v = 105\ \text{ft/sec}\quad (110\ \text{ft/sec})$$

**The answer is (B).**

Why Other Options Are Wrong

(A) This incorrect answer results when the 2 in the equation for $\sigma$ is forgotten.

(C) This incorrect answer results when the vapor pressure is not subtracted from the pressure inside the pump.

(D) This incorrect answer results when the equation for $\sigma$ is calculated with the sum of the two pressures instead of their difference.

## SOLUTION 73

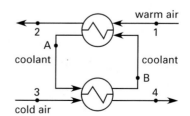

Because of symmetry, the changes in air temperature are equal and opposite.

$$T_2 - T_1 = T_3 - T_4$$
$$= \frac{Q}{(\dot{m}c_p)_{\text{air}}}$$
$$= \frac{200{,}000 \frac{\text{Btu}}{\text{hr}}}{\left(36{,}000 \frac{\text{lbm}}{\text{hr}}\right)\left(0.24 \frac{\text{Btu}}{\text{lbm-}°\text{F}}\right)}$$
$$= 23.2°\text{F}$$

The changes in coolant temperature are similarly equal and opposite. For the same heat transfer rate,

$$T_A - T_B = \frac{Q}{(\dot{m}c_p)_{\text{coolant}}}$$
$$= \frac{200{,}000 \frac{\text{Btu}}{\text{hr}}}{\left(15{,}000 \frac{\text{lbm}}{\text{hr}}\right)\left(1 \frac{\text{Btu}}{\text{lbm-}°\text{F}}\right)}$$
$$= 13.3°\text{F}$$

The average coolant temperature is

$$\overline{T}_{\text{coolant}} = \frac{T_A + T_B}{2} = \frac{T_1 + T_3}{2}$$
$$= \frac{10°\text{F} + 75°\text{F}}{2}$$
$$= 42.5°\text{F}$$

To find $T_A$ and $T_B$,

$$T_A = \overline{T}_{\text{coolant}} + \frac{T_A - T_B}{2}$$
$$= 42.5°\text{F} + \frac{13.3°\text{F}}{2} = 49.2°\text{F} \quad (49°\text{F})$$

$$T_B = \overline{T}_{\text{coolant}} - \frac{T_A - T_B}{2}$$
$$= 42.5°\text{F} - \frac{13.3°\text{F}}{2} = 35.9°\text{F} \quad (36°\text{F})$$

**The answer is (B).**

Why Other Options Are Wrong

(A) This incorrect answer results when it is assumed that $T_B = T_1$ and $T_A = T_3$.

(C) This incorrect answer results when $(T_2 - T_1)$ is used instead of instead of $(T_A - T_B)$.

(D) This incorrect answer results when $(T_2 - T_1)$ is used instead of $(T_A - T_B)$ and $T_A$ and $T_B$ are interchanged.

## SOLUTION 74

The steam quality is found using the equation

$$h = h_f + x h_{fg}$$

From the steam tables at 1800 psia,

$$h_f = 648.3 \text{ Btu/lbm}$$
$$h_{fg} = 502.1 \text{ Btu/lbm}$$

The total enthalpy in the steam, $h$, must be found. This needs to be equal to 800 Btu/lbm plus the initial enthalpy in the feedwater. At 280°F the enthalpy of the feedwater is

$$h_{\text{feed}} = 249.18 \text{ Btu/lbm}$$

The total enthalpy, then, is

$$h_{\text{total}} = h_{\text{added}} + h_{\text{feed}}$$
$$= 800 \frac{\text{Btu}}{\text{lbm}} + 249.18 \frac{\text{Btu}}{\text{lbm}}$$
$$= 1049 \text{ Btu/lbm}$$

The quality can now be found.

$$h = h_f + x h_{fg}$$
$$1049 \frac{\text{Btu}}{\text{lbm}} = 648.3 \frac{\text{Btu}}{\text{lbm}} + x \left(502.1 \frac{\text{Btu}}{\text{lbm}}\right)$$
$$x = 0.798 \quad (80\%)$$

**The answer is (C).**

Why Other Options Are Wrong

(A) This incorrect answer results when the feedwater enthalpy is subtracted from, instead of added to, the 800 Btu/lbm.

(B) This incorrect answer results when 800 Btu/lbm is used for the total energy.

(C) This incorrect answer results when 1094 is entered into the calculator instead of 1049.

## SOLUTION 75

The thermal energy provided by the combustion of natural gas is given by the equation

$$Q_{\text{comb}} = \eta \dot{m} h_{\text{HHV}}$$
$$= (0.55)\left(65{,}000 \frac{\text{lbm}}{\text{hr}}\right)\left(17{,}000 \frac{\text{Btu}}{\text{lbm}}\right)$$
$$= 607{,}750{,}000 \text{ Btu/hr}$$

The amount of thermal energy needed to increase the temperature of the natural gas from 60°F to 980°F is

$$Q_{\text{temp}} = \dot{m}c_p(T_2 - T_1)$$
$$= \left(65{,}000\ \frac{\text{lbm}}{\text{hr}}\right)\left(0.692\ \frac{\text{Btu}}{\text{lbm-°F}}\right)$$
$$\times (980°\text{F} - 60°\text{F})$$
$$= 41{,}381{,}600\ \text{Btu/hr}$$

The thermal energy available to heat the flue gas is

$$\Delta Q = Q_{\text{comb}} - Q_{\text{temp}}$$
$$= 607{,}750{,}000\ \frac{\text{Btu}}{\text{hr}} - 41{,}381{,}600\ \frac{\text{Btu}}{\text{hr}}$$
$$= 566{,}368{,}400\ \text{Btu/hr}$$

The mass flow rate of the flue gas after addition of the natural gas flow is

$$\dot{m} = \dot{m}_{\text{flue}} + \dot{m}_{\text{feed}}$$
$$= 3{,}180{,}000\ \frac{\text{lbm}}{\text{hr}} + 65{,}000\ \frac{\text{lbm}}{\text{hr}}$$
$$= 3{,}245{,}000\ \text{lbm/hr}$$

The amount of energy needed to increase the flue gas temperature is given by the equation

$$Q = \dot{m}c_p(T_{f2} - T_{f1})$$

Solving for $T_{f2}$,

$$T_{f2} = \frac{Q}{\dot{m}c_p} + T_{f1}$$
$$= \frac{566{,}368{,}400\ \frac{\text{Btu}}{\text{hr}}}{\left(3{,}245{,}000\ \frac{\text{lbm}}{\text{hr}}\right)\left(0.248\ \frac{\text{Btu}}{\text{hr-°F}}\right)} + 980°\text{F}$$
$$= 1683°\text{F}\quad(1700°\text{F})$$

**The answer is (B).**

Why Other Options Are Wrong

(A) This incorrect answer results when the 980°F temperature increase is neglected.

(C) This incorrect answer results when the energy needed to heat the natural gas and the change in mass flow rate from the addition of the natural gas are neglected.

(D) This answer results when temperatures are mistakenly converted to Rankine.

## SOLUTION 76

Statement I is false because, as the hint suggests, isothermal compression requires less power.

Statement II is also false. An isentropic process is both adiabatic and reversible, but an adiabatic process need not be reversible. So although all isentropic processes are adiabatic, not all adiabatic processes are isentropic.

Statement III is true. The work of an isentropic air compressor is

$$W = \dot{m}(h_2 - h_1) = \dot{m}c_p(T_2 - T_1)$$

For air, the specific heat is 0.244 Btu/lbm-°R and the ratio of specific heats is 1.4. The final temperature can be calculated from the initial temperature by the equation

$$T_2 = T_1\left(\frac{p_2}{p_1}\right)^{\frac{k-1}{k}}$$
$$= (530°\text{R})\left(\frac{200\ \frac{\text{lbf}}{\text{in}^2}}{14.7\ \frac{\text{lbf}}{\text{in}^2}}\right)^{\frac{1.4-1}{1.4}}$$
$$= 1120°\text{R}$$

The work can now be calculated.

$$W = \dot{m}c_p(T_2 - T_1)$$
$$= \left(1\ \frac{\text{lbm}}{\text{sec}}\right)\left(3600\ \frac{\text{sec}}{\text{hr}}\right)\left(0.24\ \frac{\text{Btu}}{\text{lbm-°R}}\right)$$
$$\times (1120°\text{R} - 530°\text{R})\left(\frac{1\ \text{hp}}{2545\ \frac{\text{Btu}}{\text{hr}}}\right)$$
$$= 200\ \text{hp}$$

Thus, only statement III is true.

**The answer is (C).**

Why Other Options Are Wrong

(A) This answer is incorrect because statement I is false.

(B) This answer is incorrect because statement II is false.

(D) This answer is incorrect because statement I is false.

## SOLUTION 77

The flow rate of the cooling water in gallons per minute can be approximated by

$$Q = \frac{H(\text{BHP})}{\left(500 \ \frac{\text{Btu-hr-min}}{°\text{F-gal}}\right) \Delta T}$$

The heat dissipation, $H$, will vary for different engines, and the value supplied by the manufacturer should be used. If it is unavailable, then for engines with water-cooled exhaust manifolds 600 Btu-hr/bhp is a safe value to use.

Usually manufacturers recommend that the temperature difference across the engine, $\Delta T$, not exceed 15°F, with 10°F being preferable. Using 10°F, the flow rate is

$$Q = \frac{\left(600 \ \frac{\text{Btu-hr}}{\text{bhp}}\right)(1100 \ \text{bhp})}{\left(500 \ \frac{\text{Btu-hr-min}}{°\text{F-gal}}\right)(10°\text{F})} = 132 \ \text{gal/min}$$

**The answer is (B).**

### Why Other Options Are Wrong

(A) This is the number of cubic feet per minute needed, not gallons per minute.

(C) This incorrect solution results from using 2200 Btu-hr/bhp for $H$ in the calculation of the required flow rate. This is the value that should be used when calculating the engine-jacket water requirement, not the lube oil's cooling water requirement.

(D) This incorrect solution results from not dividing by 10°F when calculating the flow rate.

## SOLUTION 78

The quality of the gas leaving the heater can be calculated from the formula

$$x = \frac{h_{2,\text{out}} - h_f}{h_{fg}}$$

$h_f$ and $h_{fg}$ can be found from R-12 tables at 120 psia, and equal 29.53 Btu/lbm and 58.52 Btu/lbm, respectively. The enthalpy of the gas leaving the heater, $h_{2,\text{out}}$, must be calculated.

The energy gained on the liquid side of the heater must equal the energy lost on the gas side. An energy balance on the heater gives

$$\dot{m}_1(h_{1,\text{in}} - h_{1,\text{out}}) = \dot{m}_2(h_{2,\text{out}} - h_{2,\text{in}})$$

$$h_{2,\text{out}} = h_{2,\text{in}} - \left(\frac{\dot{m}_1}{\dot{m}_2}\right)(h_{1,\text{out}} - h_{1,\text{in}})$$

The enthalpy of the saturated liquid leaving the heater, $h_{1,\text{out}}$, and the enthalpy of the superheated gas entering the heater, $h_{2,\text{in}}$, can be found from the given information and R-12 tables, and are equal to 26.49 Btu/lbm and 89.13 Btu/lbm, respectively.

The enthalpy of the saturated liquid entering the heater, $h_{1,\text{in}}$, is equal to its enthalpy entering the pump plus the enthalpy added to it by the pump. Because the liquid is known to enter the pump at 0°F and a quality of zero, its enthalpy can be found from R-12 tables and is equal to 8.25 Btu/lbm.

The enthalpy added by the pump can be calculated from the drop in pressure and the specific volume of the saturated liquid. The discharge pressure, $p_d$, is known to be 100 psia. The suction pressure, $p_s$, for the liquid at a temperature of 0°F can be found from R-12 tables and is equal to 23.87 psia. The specific volume can also be found from R-12 tables. At an estimated average pressure of 60 psia, the specific volume equals 0.0117 ft$^3$/lbm.

Neglecting velocity and potential heads, the enthalpy of the liquid entering the heater is

$$\begin{aligned} h_{1,\text{in}} &= h_{\text{pump,out}} + h_{\text{pump,added}} \\ &= h_{\text{pump,in}} + v(p_d - p_s) \\ &= 8.25 \ \frac{\text{Btu}}{\text{lbm}} + \left(0.0117 \ \frac{\text{ft}^3}{\text{lbm}}\right) \\ &\quad \times \left(100 \ \frac{\text{lbf}}{\text{in}^2} - 23.87 \ \frac{\text{lbf}}{\text{in}^2}\right)\left(\frac{1 \ \text{Btu}}{778 \ \text{ft-lbf}}\right) \\ &\quad \times \left(12 \ \frac{\text{in}}{\text{ft}}\right)^2 \\ &= 8.41 \ \text{Btu/lbm} \end{aligned}$$

The enthalpy of the gas exiting the heater can now be calculated.

$$\begin{aligned} h_{2,\text{out}} &= h_{2,\text{in}} - \left(\frac{\dot{m}_1}{\dot{m}_2}\right)(h_{1,\text{out}} - h_{1,\text{in}}) \\ &= 89.13 \ \frac{\text{Btu}}{\text{lbm}} - \left(\frac{1 \ \frac{\text{lbm}}{\text{sec}}}{0.5 \ \frac{\text{lbm}}{\text{sec}}}\right) \\ &\quad \times \left(26.49 \ \frac{\text{Btu}}{\text{lbm}} - 8.41 \ \frac{\text{Btu}}{\text{lbm}}\right) \\ &= 53.0 \ \text{Btu/lbm} \end{aligned}$$

The quality of the gas exiting the heater is

$$x = \frac{h_{2,\text{out}} - h_f}{h_{fg}}$$

$$= \frac{53.0 \frac{\text{Btu}}{\text{lbm}} - 29.53 \frac{\text{Btu}}{\text{lbm}}}{58.52 \frac{\text{Btu}}{\text{lbm}}}$$

$$= 0.401 \quad (40\%)$$

**The answer is (B).**

Why Other Options Are Wrong

(A) This incorrect answer results when the R-12 tables are misread and $h_g$ is used instead of $h_{fg}$.

(C) This incorrect answer results when the conversion from Btu to ft-lbf is neglected.

(D) This incorrect answer results when the mass flow rates are switched.

# CODES AND STANDARDS

## SOLUTION 79

Since the viscosity of the oil is temperature dependent, a trial-and-error solution method is required. Design charts from Raimondi and Boyd assume, however, that the oil's viscosity is constant as it passes through the bearing. (These charts can be found in *Mechanical Engineering Design* by Shigley and Mischke.)

Therefore, the average of the oil inlet and exit temperatures is used to determine the viscosity. Begin by assuming the temperature rise will be 60°F. The average oil temperature is

$$T_{\text{ave}} = \frac{120°\text{F} + 180°\text{F}}{2} = 150°\text{F}$$

From a viscosity-temperature chart, the viscosity ($\mu$) of SAE 30 oil at 150°F is $3.6 \times 10^{-6}$ reyns.

The temperature rise for an oil with this viscosity is found using the equation

$$\Delta T_F = \left( \frac{\left(0.103 \frac{\text{in}^2\text{-°F}}{\text{lbf}}\right) P}{1 - \left(\frac{1}{2}\right)\left(\frac{Q_s}{q}\right)} \right) \left( \frac{\left(\frac{r}{c}\right) f}{\frac{Q}{rcnL}} \right)$$

The friction variable $(r/c)f$, the flow rate $Q_s/q$, and the flow variable $Q/rcnL$ are found using charts where these variables are plotted against the Sommerfeld number for various $L/d$ ratios. The Sommerfeld number is

$$S = \left(\frac{r}{c_r}\right)^2 \left(\frac{\mu n}{P}\right)$$

The load per unit of projected bearing, $P$, is

$$P = \frac{W}{2rL} = \frac{495 \text{ lbf}}{(2)(0.75 \text{ in})(1.5 \text{ in})} = 220 \text{ lbf/in}^2$$

The $L/d$ ratio is equal to 1, since the bearing length, $L$, is equal to the diameter.

$$S = \left(\frac{r}{c_r}\right)^2 \left(\frac{\mu n}{P}\right)$$

$$= \left(\left(\frac{0.75 \text{ in}}{2 \text{ mils}}\right)\left(1000 \frac{\text{mils}}{\text{in}}\right)\right)^2$$

$$\times \frac{(3.6 \times 10^{-6} \text{ reyn})\left(1 \frac{\frac{\text{lbf-sec}}{\text{in}^2}}{\text{reyn}}\right) \times \left(3600 \frac{\text{rev}}{\text{min}}\right)\left(\frac{1 \text{ min}}{60 \text{ sec}}\right)}{220 \frac{\text{lbf}}{\text{in}^2}}$$

$$= 0.14$$

From the charts,

$$\left(\frac{r}{c}\right) f = 3.6$$

$$\frac{Q}{rcnL} = 4.3$$

$$\frac{Q_s}{Q} = 0.65$$

Using the earlier equation, the temperature rise is

$$\Delta T_F = \left( \frac{\left(0.103 \frac{\text{in}^2\text{-°F}}{\text{lbf}}\right)\left(220 \frac{\text{lbf}}{\text{in}^2}\right)}{1 - \left(\frac{1}{2}\right)(0.65)} \right)\left(\frac{3.6}{4.3}\right)$$

$$= 28°\text{F}$$

The average temperature based on this temperature rise is

$$T_{\text{ave}} = T_{\text{in}} + \frac{\Delta T_F}{2} = 120°\text{F} + \frac{28°\text{F}}{2} = 134°\text{F}$$

Now the ordered pair (134°F, 3.6 × 10⁻⁶ reyns) can be graphed on the viscosity-temperature chart. Graphing this point and labeling it point A, it can be seen to be below the SAE 30 line, and thus the viscosity used was too low. Choosing a viscosity above the SAE 30 line—say, 6.0 × 10⁻⁶ reyns—will give the following temperature rise.

$$S = \left(\left(\frac{0.75 \text{ in}}{2 \text{ mils}}\right)\left(1000 \frac{\text{mils}}{\text{in}}\right)\right)^2$$

$$\times \frac{(6 \times 10^{-6} \text{ reyn})\left(1 \frac{\frac{\text{lbf-sec}}{\text{in}^2}}{\text{reyn}}\right)}{220 \frac{\text{lbf}}{\text{in}^2}} \times \left(3600 \frac{\text{rev}}{\text{min}}\right)\left(\frac{1 \text{ min}}{60 \text{ sec}}\right)$$

$$= 0.23$$

From the charts,

$$\left(\frac{r}{c}\right)f = 5.5$$

$$\frac{Q}{rcnL} = 4.05$$

$$\frac{Q_s}{Q} = 0.53$$

The temperature rise is

$$\Delta T_F = \left(\frac{\left(0.103 \frac{\text{in}^2\text{-}°F}{\text{lbf}}\right)\left(220 \frac{\text{lbf}}{\text{in}^2}\right)}{1 - \left(\frac{1}{2}\right)(0.53)}\right)\left(\frac{5.5}{4.05}\right)$$

$$= 42°F$$

The average temperature based on this temperature rise is

$$T_{\text{ave}} = T_{\text{in}} + \frac{\Delta T_F}{2} = 120°F + \frac{42°F}{2} = 141°F$$

Now the ordered pair (141°F, 6.0 × 10⁻⁶ reyns) can be graphed on the viscosity-temperature chart and labeled as point B. Draw a line between points A and B and find the intersection of this line with the SAE 30 line. This point gives the correct oil viscosity and average temperature for this bearing under these conditions. The intersection point is (138°F, 4.8 × 10⁻⁶ reyns).

The temperature rise is

$$T_{\text{ave}} = \frac{T_{\text{in}} + (T_{\text{in}} + \Delta T)}{2} = T_{\text{in}} + \frac{\Delta T}{2}$$

$$\Delta T = 2(T_{\text{ave}} - T_{\text{in}}) = (2)(138°F - 120°F) = 36°F$$

**The answer is (C).**

Why Other Options Are Wrong

(A) This incorrect answer results when the factor of two is neglected.

(B) This incorrect answer results when the $(r/c)f$ and $Q/rcnL$ values are switched when calculating $T_F$ for point B.

(D) This incorrect solution results when the temperature $T_{\text{ave}}$ is misread from the chart as 148°F.

## SOLUTION 80

The remaining corrosion allowance (RCA) is found using

$$\text{RCA} = t_{\text{actual}} - t_{\text{min}}$$

ASME code B31.3 says that either of the following equations can be used to determine the minimum thickness, $t_{\text{min}}$.

$$t_{\text{min}} = \frac{Pd}{2(SE + PY)}$$

$$t_{\text{min}} = \frac{Pd}{2SE}$$

$Y$ is a coefficient obtained from a table. The second equation is the one more often used to determine the minimum thickness. Because the stress value is usually so much larger than the internal pressure, the $PY$ term in the denominator of the first equation is frequently negligible.

From schedule-40 pipe tables, the outside diameter of 3 in pipe is 3.5 in. From the ASME *Boiler and Pressure Vessel Code,* Sec. II, the stress value, $S$, for SA-106 grade B pipe at 600°F is 17.1 ksi. The quality factor, $E$, for seamless pipe is 1. The minimum thickness is

$$t_{\text{min}} = \frac{Pd}{2SE}$$

$$= \frac{\left(250 \frac{\text{lbf}}{\text{in}^2}\right)(3.5 \text{ in})}{(2)\left(17{,}100 \frac{\text{lbf}}{\text{in}^2}\right)(1)}$$

$$= 0.0256 \text{ in}$$

The RCA is

$$\text{RCA} = 0.175 \text{ in} - 0.0256 \text{ in} = 0.149 \text{ in} \quad (0.15 \text{ in})$$

**The answer is (B).**

## Why Other Options Are Wrong

(A) This incorrect solution results from neglecting the 2 in the denominator.

(C) This solution results from using a wrong diameter of 3 in, a wrong stress value of 20,000 lbf/in², and a factor of safety of 1.67, which is not needed because ASME Sec. II gives the corrected stress value.

(D) This incorrect solution results from using the original pipe thickness for the actual thickness.

## SOLUTION 81

In the ASME *Boiler and Pressure Vessel Code*, Sec. VIII, Div. 1, the reinforcing area of the shell plus the reinforcing area of the nozzle must equal at least the required reinforcement area.

$$A_s + A_n \geq A_r$$

Because the nozzle abuts the vessel wall, and using a correction factor of 1.0, the required reinforcement area is found using

$$A_r = dt_r$$

$d$ is the finished diameter of the circular opening in the shell, and $t_r$ is the required thickness of the shell based on the circumferential stress.

The required thickness of the shell is found using

$$t_{\text{req}} = \frac{PR}{SE - 0.6P}$$

The allowable stress, $S$, of SA-106 grade B material from ASME Sec. II is 17.1 ksi, and the efficiency, $E$, of the joint in a seamless shell is 1.

$$t_{\text{req}} = \frac{PR}{SE - 0.6P}$$

$$= \frac{\left(250 \, \frac{\text{lbf}}{\text{in}^2}\right)(34 \text{ in})}{\left(17{,}100 \, \frac{\text{lbf}}{\text{in}^2}\right)(1.0) - (0.6)\left(250 \, \frac{\text{lbf}}{\text{in}^2}\right)}$$

$$= 0.501 \text{ in}$$

Using pipe size tables, the required reinforcement area is found to be

$$A_{\text{req}} = (4.5 \text{ in})(0.501 \text{ in}) = 2.25 \text{ in}^2$$

The area available in the shell is

$$A_s = d(t - t_{\text{req}}) = (4.5 \text{ in})(0.625 \text{ in} - 0.501 \text{ in})$$

$$= 0.558 \text{ in}^2$$

$$A_s + A_n = 0.558 \text{ in}^2 + 0.518 \text{ in}^2 = 1.08 \text{ in}^2$$

Since this is less than the required area, the amount of additional reinforcement needed is

$$A_{\text{add}} = 2.25 \text{ in}^2 - 1.08 \text{ in}^2 = 1.17 \text{ in}^2 \quad (1.2 \text{ in}^2)$$

**The answer is (B).**

## Why Other Options Are Wrong

(A) This incorrect answer results when the wrong diameter for the nozzle is used.

(C) This incorrect answer results when an incorrect efficiency of 0.6 is used to calculate $t_{\text{req}}$.

(D) This incorrect answer results when the vessel's diameter is used to calculate $t_{\text{req}}$.

## SOLUTION 82

Section I of the ASME *Boiler and Pressure Vessel Code* states that the relieving capacities of steam relief safety valves are given by

$$W_{\text{th}} = 51.5 Ap$$

The theoretical flow, $W_{\text{th}}$, is in lbm/hr; the nozzle throat area, $A$, is in square inches; and the pressure, $p$, for 3% accumulation, is

$$p = 1.03(\text{set pressure}) + 14.7 \, \frac{\text{lbf}}{\text{in}^2}$$

$$= (1.03)\left(1400 \, \frac{\text{lbf}}{\text{in}^2}\right) + 14.7 \, \frac{\text{lbf}}{\text{in}^2}$$

$$= 1456.7 \text{ lbf/in}^2 \quad [\text{pressure absolute}]$$

Note that the ASME code says to use for $p$ either the preceding equation or the one following, whichever is greater.

$$p = \text{set pressure} + 2 + 14.7 \, \frac{\text{lbf}}{\text{in}^2}$$

The latter equation, however, will only be greater for set pressures equal to or less than 67 psig.

Substituting into the steam flow equation and solving for $A$ gives

$$W_{\text{th}} = 51.5 Ap$$

$$200{,}000 \, \frac{\text{lbm}}{\text{hr}} = \left(51.5 \, \frac{\text{lbm}}{\text{hr-lbf}}\right) A \left(1456.7 \, \frac{\text{lbf}}{\text{in}^2}\right)$$

$$A = 2.7 \text{ in}^2$$

**The answer is (A).**

## Why Other Options Are Wrong

(B) This incorrect answer results from using an older ASME code giving the relieving capacity as $W_T = 45AP$.

(C) This incorrect answer results from misreading 51.5 as 15.5.

(D) This incorrect answer results from using 140 psig in the calculation instead of 1400 psig.

## SOLUTION 83

API 682 states that the injection flow rate for piping plan 12 can be determined using the equation

$$Q_{inj} = \frac{14.35 P}{(SG)(dT)c_p}$$

When using this equation, this standard states that a design factor of least two should be applied to the flow rate.

To make the injection rate equation dimensionally correct, the constant must have the following units.

$$Q_{inj} = \left(14.35 \, \frac{\text{L·s}}{\text{kg·min}}\right) \left(\frac{P}{(SG)(dT)c_p}\right)$$

To use this equation, the heat generation at the seal forces, $P$, must be in kilowatts; the differential temperature, $dT$, between the pump and the desired seal chamber temperature must be in degrees Celsius; and the specific heat, $c_p$, of the injected fluid must be in kJ/kg·°C.

API 682 states that for good performance the temperature rise should be between 2.8°C and 5.6°C, inclusive. Therefore, the minimum allowable injection flow rate is the one that gives a value of 5.6°C for $dT$.

For water,

$$SG = 1.0$$
$$c_p = 4.18 \text{ kJ/kg·°C}$$

Before allowing for the design factor, the minimum allowable injection flow rate is

$$Q_{inj} = \left(14.35 \, \frac{\text{L·s}}{\text{kg·min}}\right) \left( \frac{(1.5 \text{ kW})\left(1 \, \frac{\text{s}}{\text{kW}}\frac{\text{kJ}}{}\right)}{(1.0)(5.6°\text{C})\left(4.18 \, \frac{\text{kJ}}{\text{kg·°C}}\right)} \right)$$

$$= 0.92 \text{ L/min}$$

The design factor of two makes the minimum allowable injection flow rate approximately 1.8 L/min.

**The answer is (B).**

## Why Other Options Are Wrong

(A) This incorrect answer results when the design factor of two is neglected.

(C) This incorrect answer results when a temperature rise of 2.8°C is used instead of 5.6°C.

(D) This incorrect answer results when the numerical value for the specific heat in Btu/lbm-°F is mistakenly used.

## SOLUTION 84

From the ASME *Boiler and Pressure Vessel Code*, Sec. VIII, Div. 1, the thickness is calculated using the equation

$$t = d\sqrt{\frac{Cp}{SE}}$$

Since a blind flange is normally seamless, the joint efficiency, $E$, is equal to 1.

$$t = (2 \text{ ft})\left(12 \, \frac{\text{in}}{\text{ft}}\right) \sqrt{\frac{(0.17)\left(200 \, \frac{\text{lbf}}{\text{in}^2}\right)}{\left(13{,}000 \, \frac{\text{lbf}}{\text{in}^2}\right)(1)}}$$

$$= 1.2 \text{ in}$$

**The answer is (B).**

## Why Other Options Are Wrong

(A) This incorrect answer results when the diameter is not converted to inches.

(C) This incorrect answer results when the attachment factor is ignored.

(D) This incorrect answer results when the diameter is not converted to inches and the design pressure and allowable stress are interchanged.

## SOLUTION 85

From the ASME *Boiler and Pressure Vessel Code*, Sec. VIII, Div. 1, the minimum thickness is calculated using the equation

$$t_{min} = d\sqrt{\frac{Cp}{SE}}$$

$$= (3 \text{ ft})\left(12 \, \frac{\text{in}}{\text{ft}}\right) \sqrt{\frac{(0.20)\left(150 \, \frac{\text{lbf}}{\text{in}^2}\right)}{\left(17{,}500 \, \frac{\text{lbf}}{\text{in}^2}\right)(1)}}$$

$$= 1.5 \text{ in}$$

The thickness of the head can then be calculated using the equation

$$t_{head} = t_{min} + (CR)L$$
$$= 1.5 \text{ in} + \left(8 \; \frac{\text{mil}}{\text{yr}}\right)\left(\frac{1 \text{ in}}{1000 \text{ mil}}\right)(30 \text{ yr})$$
$$= 1.74 \text{ in} \quad (1.7 \text{ in})$$

**The answer is (C).**

Why Other Options Are Wrong

(A) The minimum design thickness must be greater than just the corrosion amount of 0.24 in over 30 years or the head will not be able to withstand the pressure.

(B) This incorrect answer results when the diameter is not converted to inches.

(D) This incorrect answer results when the attachment factor is ignored.